The series "Advances in Intelligent Systems and Computing" contains publications on theory, applications, and design methods of Intelligent Systems and Intelligent Computing. Virtually all disciplines such as engineering, natural sciences, computer and information science, ICT, economics, business, e-commerce, environment, healthcare, life science are covered. The list of topics spans all the areas of modern intelligent systems and computing such as: computational intelligence, soft computing including neural networks, fuzzy systems, evolutionary computing and the fusion of these paradigms, social intelligence, ambient intelligence, computational neuroscience, artificial life, virtual worlds and society, cognitive science and systems, Perception and Vision, DNA and immune based systems, self-organizing and adaptive systems, e-Learning and teaching, human-centered and human-centric computing, recommender systems, intelligent control, robotics and mechatronics including human-machine teaming, knowledge-based paradigms, learning paradigms, machine ethics, intelligent data analysis, knowledge management, intelligent agents, intelligent decision making and support, intelligent network security, trust management, interactive entertainment, Web intelligence and multimedia.

The publications within "Advances in Intelligent Systems and Computing" are primarily proceedings of important conferences, symposia and congresses. They cover significant recent developments in the field, both of a foundational and applicable character. An important characteristic feature of the series is the short publication time and world-wide distribution. This permits a rapid and broad dissemination of research results.

More information about this series at http://www.springer.com/series/11156

Jessie Chen
Editor

Advances in Human Factors in Robots and Unmanned Systems

Proceedings of the AHFE 2018 International
Conference on Human Factors in Robots
and Unmanned Systems, July 21–25, 2018,
Loews Sapphire Falls Resort at Universal Studios,
Orlando, Florida, USA

 Springer

Editor
Jessie Chen
U.S. Army Research Laboratory
Orlando, FL, USA

ISSN 2194-5357 ISSN 2194-5365 (electronic)
Advances in Intelligent Systems and Computing
ISBN 978-3-319-94345-9 ISBN 978-3-319-94346-6 (eBook)
https://doi.org/10.1007/978-3-319-94346-6

Library of Congress Control Number: 2018947357

This Springer imprint is published by the registered company Springer International Publishing AG
part of Springer Nature
The registered company address is: Gewerbestrasse 11, 6330 Cham, Switzerland

Advances in Human Factors and Ergonomics 2018

AHFE 2018 Series Editors

Tareq Z. Ahram, Florida, USA
Waldemar Karwowski, Florida, USA

9th International Conference on Applied Human Factors and Ergonomics and the Affiliated Conferences

Proceedings of the AHFE 2018 International Conferences on Human Factors in Robots and Unmanned Systems, held on July 21–25, 2018, in Loews Sapphire Falls Resort at Universal Studios, Orlando, Florida, USA

Advances in Affective and Pleasurable Design	Shuichi Fukuda
Advances in Neuroergonomics and Cognitive Engineering	Hasan Ayaz and Lukasz Mazur
Advances in Design for Inclusion	Giuseppe Di Bucchianico
Advances in Ergonomics in Design	Francisco Rebelo and Marcelo M. Soares
Advances in Human Error, Reliability, Resilience, and Performance	Ronald L. Boring
Advances in Human Factors and Ergonomics in Healthcare and Medical Devices	Nancy J. Lightner
Advances in Human Factors in Simulation and Modeling	Daniel N. Cassenti
Advances in Human Factors and Systems Interaction	Isabel L. Nunes
Advances in Human Factors in Cybersecurity	Tareq Z. Ahram and Denise Nicholson
Advances in Human Factors, Business Management and Society	Jussi Ilari Kantola, Salman Nazir and Tibor Barath
Advances in Human Factors in Robots and Unmanned Systems	Jessie Chen
Advances in Human Factors in Training, Education, and Learning Sciences	Salman Nazir, Anna-Maria Teperi and Aleksandra Polak-Sopińska
Advances in Human Aspects of Transportation	Neville Stanton

(continued)

(continued)

Advances in Artificial Intelligence, Software and Systems Engineering	*Tareq Z. Ahram*
Advances in Human Factors, Sustainable Urban Planning and Infrastructure	*Jerzy Charytonowicz and Christianne Falcão*
Advances in Physical Ergonomics & Human Factors	*Ravindra S. Goonetilleke and Waldemar Karwowski*
Advances in Interdisciplinary Practice in Industrial Design	*WonJoon Chung and Cliff Sungsoo Shin*
Advances in Safety Management and Human Factors	*Pedro Miguel Ferreira Martins Arezes*
Advances in Social and Occupational Ergonomics	*Richard H. M. Goossens*
Advances in Manufacturing, Production Management and Process Control	*Waldemar Karwowski, Stefan Trzcielinski, Beata Mrugalska, Massimo Di Nicolantonio and Emilio Rossi*
Advances in Usability, User Experience and Assistive Technology	*Tareq Z. Ahram and Christianne Falcão*
Advances in Human Factors in Wearable Technologies and Game Design	*Tareq Z. Ahram*
Advances in Human Factors in Communication of Design	*Amic G. Ho*

Preface

Researchers are conducting cutting-edge investigations in the area of unmanned systems to inform and improve how humans interact with robotic platforms. Many of the efforts are focused on refining the underlying algorithms that define system operation and on revolutionizing the design of human–system interfaces. The multifaceted goals of this research are to improve ease of use, learnability, suitability, and human–system performance, which in turn will reduce the number of personnel hours and dedicated resources necessary to train, operate, and maintain the systems. As our dependence on unmanned systems grows along with the desire to reduce the manpower needed to operate them across both the military and commercial sectors, it becomes increasingly critical that system designs are safe, efficient, and effective. Optimizing human–robot interaction and reducing cognitive workload at the user interface require research emphasis to understand what information the operator requires, when they require it, and in what form it should be presented so they can intervene and take control of unmanned platforms when it is required. With a reduction in manpower, each individual's role in system operation becomes even more important to the overall success of the mission or task at hand. Researchers are developing theories as well as prototype user interfaces to understand how best to support human–system interaction in complex operational environments. Because humans tend to be the most flexible and integral part of unmanned systems, the human factors and unmanned systems' focus considers the role of the human early in the design and development process in order to facilitate the design of effective human–system interaction and teaming.

This book will prove useful to a variety of professionals, researchers, and students in the broad field of robotics and unmanned systems who are interested in the design of multisensory user interfaces (auditory, visual, and haptic), user-centered design, and task–function allocation when using artificial intelligence/automation to offset cognitive workload for the human operator. We hope this book is informative, but even more than that it is thought-provoking. We hope it provides inspiration, leading the reader to formulate new, innovative research questions, applications, and potential solutions for creating effective human–system interaction and teaming with robots and unmanned systems. Two sections presented in this book:

   I. Human Interaction with Unmanned and Autonomous Systems
  II. Human-Robot Collaborations and Interactions

Each section contains research papers that have been reviewed by members of the International Editorial Board. Our sincere thanks and appreciation to the Board members as listed below:

Michael Barnes, USA
Paulo Bonato, USA
Gloria Calhoun, USA
Reece Clothier, Australia
Nancy Cooke, USA
Linda Elliott, USA
Daniel Ferris, USA
Janusz Fraczek, Poland
Joseph W. Geeseman, USA
Jonathan Gratch, USA
Susan Hill, USA
Eric Holder, USA
Ming Hou, Canada
Chris Johnson, UK
Troy Kelley, USA
Michael LaFiandra, USA
Joseph Lyons, USA
Kelly Neville, USA
Jacob N. Norris, USA
Jose L. Pons, Spain
Charlene Stokes, USA
Peter Stütz, Germany
Redha Taiar, France
Jeffrey Thomas, USA
Anna Trujillo, USA
Anthony Tvaryanas, USA
Herman Van der Kooij, the Netherlands
Harald Widlroither, Germany
Huiyu Zhou, UK

July 2018   Jessie Chen

Contents

Human Interaction with Unmanned and Autonomous Systems

Operator Trust Function for Predicted Drone Arrival

Anna C. Trujillo[⊠]

NASA Langley Research Center, MS 152, Hampton, VA, USA
anna.c.trujillo@nasa.gov

Abstract. To realize the full benefit from autonomy, systems will have to react to unknown events and uncertain dynamic environments. The resulting number of behaviors is essentially infinite; thus, the system is effectively non-deterministic but an operator needs to understand and trust the actions of the autonomous vehicles. This research began to tackle non-deterministic systems and trust by beginning to develop a user trust function based on intent information displayed and the prescribed bounds on allowable behaviors/actions of the non-deterministic system. Linear regression shows promise on being able to predict a person's confidence of the machine's prediction. Linear regression techniques indicated that subject characteristics, scenario difficulty, the experience with the system, and confidence earlier in the scenario account for approximately 60% of the variation in confidence ratings. This paper details the specifics of the liner regression model – essentially a trust function – for predicting a person's confidence.

Keywords: Trust function · Non-deterministic system · Linear regression
Autonomy · Confidence rating

1 Introduction

A primary goal is for public and civil operators is to have one person managing several vehicles with different mission goals. To realize the full benefit from autonomy, these systems will have to react to unknown events and uncertain dynamic environments. The resulting number of behaviors is essentially infinite; thus, the system is effectively non-deterministic. So, rather than verify that an autonomous agent provides the correct answer in all cases, an impossibility with a non-deterministic system, instead determine whether verifying a defined solution space (i.e., bounds on system behavior) is feasible.

An operator overseeing a group of autonomous vehicles is a direct application of this approach. The operator needs to understand and trust the actions of autonomous vehicles. Achieving trust will become even more difficult and complicated if vehicles are able to make effectively non-deterministic decisions. An operator may learn to not trust or have confidence in such a vehicle if he is unable to understand why an autonomous vehicle is taking a particular action, which could result in system-wide failures and limited system performance due to the operator overriding the autonomous vehicle's protocols. Conversely, if the operator relies on autonomy in excess of its behavioral bounds, he may lose situation awareness of the system as a whole with consequences

J. Chen (Ed.): AHFE 2018, AISC 784, pp. 3–14, 2019.
https://doi.org/10.1007/978-3-319-94346-6_1

ranging from suboptimal system performance to potentially catastrophic for conditions falling outside the autonomous vehicle's behavior boundary constraints. This research began to tackle these problems by trying to quantify the solution space non-deterministic systems inhabit as a function of the mission and then informing the operator of this solution space to foster trust and increase efficiency of the system as a whole.

The overall objectives were two-fold – (1) verifying bounds for non-deterministic decisions and (2) fostering trust in the actions taken by autonomous agents by making their decision process transparent to the operator. Objectives for this initial experiment were to (a) develop a non-deterministic system that operates within known bounds, (b) develop a display that shows possible outcomes from the current state, and (c) begin to develop a user trust function based on intent information displayed and the pre-scribed bounds on allowable behaviors/actions of the non-deterministic system. The autonomous agent had responsibility for mission performance that entailed trajectory planning and replanning in response to unanticipated events such as weather, other aircraft, etc. without outside operator supervision. The system behavior bounds were dependent on vehicle internal state self-knowledge, external environment represented by sensor data, and associated uncertainty. To have known and hence prescribed bounds, a modified Chua's circuit [1, 2] initially modeled the non-deterministic system. Autonomous agent intent information to display for user trust function initial devel-opment involved a drone arriving at its next waypoint at a specified latitude, longitude, altitude and time – a 4D trajectory.

2 Background

Research is just beginning on displaying possible outcomes and decision making (for example, see [3–5]) and previous research on trust focused on subjective measures (for example, see [6, 7]) rather than objective measures. Various models and ques-tionnaires to define trust have been developed. For example, Hoff and Bashir [8] developed a three-layered framework for trust. This framework contains dispositional trust, which includes personality traits, situational trust, which encompasses workload, task difficulty and self-confidence, and learned trust, which includes experience with the system, knowledge of system performance and transparency. Others have found that personality traits may not significantly affect trust; instead, trust was more affected by the autonomous agent characteristics and the task characteristics [9].

The Army Research Laboratory has focused recently on agent transparency effects on operator trust [10–13]. They have found that increasing transparency increased operator's trust [10] but this increase is limited [11]. With this added information on agent transparency, increases in operator workload may occur due to additional information the operator must pay attention to; however, recent research indicated that added transparency information did not significantly increase workload [14, 15].

The Army Research Laboratory situation awareness-based agent transparency (SAT) model [12] mirrors Endsley's situation awareness model [16, 17] in that it has three levels. The first level consists of basic information such as purpose, process, and current performance and status. The second level consists of rationale or the agent's reasoning process which may include environmental and other constraints. The third level consists of outcomes and includes projections of future outcomes, uncertainty, likelihood of success, and performance history.

Considering Chen's, et al. SAT model [12] and trust questionnaires [7–9], operator personality traits, autonomous agent history, and the task itself, this initial experiment looked to develop an objective trust function. This trust function could then be used to predetermine the needed information to provide to an operator to ensure trust is maintained among all agents and to maintain trust during an operation by changing the information provided to the operator based on signal variations. This will allow for an optimal system by ensuring that a lack of trust does not lead to an operator overriding the system [18] and that excessive trust does not lead to a lack of operator situation awareness [19].

This research looked to define objective measures that estimate user trust in a system. First, a non-deterministic system that operates within known bounds was designed and is described in Sects. 3.1 and 3.2. Second, a simple display indicating the predicted state of a drone arriving at a 4D gate was designed and is described in Sect. 3.3. Lastly, a human-in-the-loop experiment, described in Sect. 4, collected user trust of the system and this data was used to develop an initial trust function described in Sect. 5.

3 Setup

3.1 Chua's Circuit

A Chua's circuit was used to mimic a non-deterministic system that operates within known bounds. Chua's circuits have values that are theoretically prov-able to fall within a defined range [1, 2, 20]. A basic Chua's circuit is shown in Fig. 1. The Chua diode consisted of resistors 1 thru 6. The gyrator, or inductor, consisted of resistors 7 thru 10 and the capacitor C. The values for these resistors, inductor, and capacitor were from [20]. C_1, C_2, and R depended on the scenario. Because four values were needed, the Chua's circuit was run twice with the same C_1 and C_2 values but with

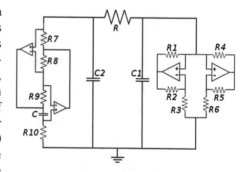

Fig. 1. Diagram of Chua's circuit (figure from http://www.chuacircuits.com/matlabsim.php).

different R values. C_1 ranged from 8 nF (nanofarad) to 10 nF, C_2 ranged from 80 nF to 100 nF, and R ranged from 1800 Ohms to 2100 Ohms.

3.2 Flight Paths

The drone followed four types of trajectory paths. The first was constant velocity latitude and longitude change with constant altitude. The second was constant velocity

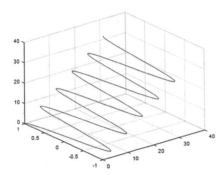

(a) Boustrophedon flight path with constantly changing altitude.

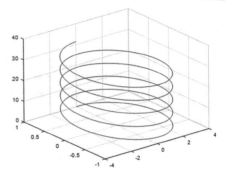

(b) Constantly ascending or descending helix flight path.

Fig. 2. Boustrophedon and helix flight paths.

latitude and longitude change with constant changing altitude (ascending or descending) (Fig. 2). The third consisted of a smooth boustrophedon pattern with constant changing altitude (ascending or descending) (a) and the last trajectory was a constantly ascending or descending helix (b).

The Chua's circuit simulated the drone's deviation from the flight path, specifically longitude, latitude, altitude, and time. The absolute maximum or bounds of the Chua's circuit to the prescribed flight paths had three variation levels – low, medium, and high (normalized latitude, longitude, altitude = ±10, ±12, ±14 and time = ±5, ±7, ±9 respectively) – and one scenario that actually exceeded the bounds at the end of the run – Exceed (latitude, altitude = ±14, longitude = ±16 and time = ±7). Figure 3 depicts an example of the prescribed path and deviations from the prescribed path as generated from the Chua's circuit.

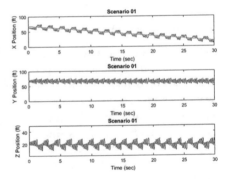

Fig. 3. Example constant velocity latitude and longitude change with constant altitude flight path with low variation.

3.3 Displays

Each subject saw several different displays – (1) trajectory display only, (2) trajectory display plus predicted deviation from prescribed gate, (3) previous plus deviation bars, (4) previous plus predicted deviation values from prescribed gate, and (5) previous plus likelihood of predicted deviation values from prescribed gate.

Trajectory Display. The trajectory display showed the current position of the drone relative to its prescribed latitude, longitude, and altitude positions (left side of Fig. 4).

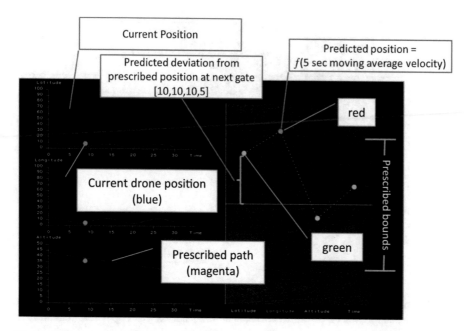

Fig. 4. Trajectory display plus predicted deviation from prescribed gate. Predicted position deviation is normalized. Prescribed bounds area is a green shade while outside the prescribed bounds area is a red shade. Predicted position dot and its text is same color as the area it is in.

Trajectory Display Plus Predicted Deviation from Prescribed Gate. This display added a panel to show the weighted average predicted deviation from the prescribed gate which was to be reached at the end of the run (right side of Fig. 4). The green area indicated the acceptable prescribed bounds which were ±10 units for latitude, longitude and altitude and ±5 s for time. The red areas were exceedances of these bounds. The predicted position was a function of the 5 s moving average velocity of latitude, longitude, altitude, and time. If a value was predicted to exceed the bounds when the drone was to reach the gate, the dot and its associated text were colored red otherwise they were green. The light dotted white line between the values was there to aid the subject in lining up the dot location with its heading text.

Trajectory Display Plus Predicted Deviation from Prescribed Gate with Deviation Bars. This display added deviation bars that indicated the range of possible values (Fig. 5). The deviations were a function of average velocity and velocity of the 4D gate. The high and low values did not have to be equal.

Trajectory Display Plus Predicted Deviation from Prescribed Gate with Deviation Bars and Predicted Deviation Values. This display added predicted deviation values for the predicted value and its high and low value (Fig. 5). As with the dots being color coded, the values were the same color as where the value resided. So, if the value was within the prescribed bounds, the number was green; otherwise the number was red.

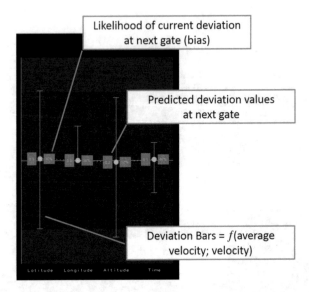

Fig. 5. Predicted deviation from prescribed gate with deviation bars, predicted deviation values at next gate, and likelihood of current deviation at next gate. As with Fig. 4, text color matches the area the value is in.

Trajectory Display Plus Predicted Deviation from Prescribed Gate with Deviation Bars, Predicted Deviation and Likelihood of Predicted Deviation Values. The final display added the likelihood of the predicted deviation values (Fig. 5). As before, the percentages were the same color as the predicted deviation values.

3.4 Confidence Rating Question

At three points during each data run (at 10 s, 20 s, and at the end of the run), the scenario paused so that subjects could answer "How confident are you that the drone would reach its gate?" This question was on a scale of 0, no confidence that the drone would reach its gate within the prescribed boundaries, and 100, absolutely sure that the drone would reach its gate within the prescribed boundaries.

3.5 Electronic NASA-TLX Questionnaire

At the end of each run, subjects completed an electronic version of the NASA Task Load Index (NASA-TLX) [21]. This questionnaire was always the last set of questions presented at the end of the 30 s run.

4 Procedure

Each subject had 15 data runs – 3 with each of the displays described in Sect. 3.3. Each data run lasted for 30 s and at the 10 s, 20 s, and 30 s points, the scenario paused so that the subject could answer the questions described in Sect. 3.4. At the end of the run, subjects also completed an electronic version of the NASA-TLX. At a change of display, subjects had 2 practice runs that behaved just like the data runs.

5   Results for End Confidence Rating

Data was analyzed using IBM® SPSS®[1] V24 automatic linear regression techniques. Significance was taken at $p \leq 0.05$.

The linear regression to predict confidence rating at the end of the run had an accuracy of 62%. The significant factors are earlier confidence ratings during the run, subject personality, path deviation level, when the run occurred, and an intercept of 36 (Eq. 1).

$$CR_{end} = 36 + 0.24CR_{20\,s} + 0.21CR_{10\,s} + subject_{personality} + path_{deviation} + run_n \quad (1)$$

where

$$CR = \text{confidence rating at end, 20 s, or 10 s}$$

$$subject_{personality} = \begin{cases} 15 & \text{if } low\,frustration \text{ and } good\,performance \\ 9 & \text{if } mid\,frustration \text{ and } mid\,performance \\ 0 & \text{if } high\,frustration \text{ and } poor\,performance \end{cases}$$

$$path_{deviation} = \begin{cases} 0 & \text{if } deviation \neq high \\ -9 & \text{if } deviation = high \end{cases}$$

$$run_n = \begin{cases} 6 & \text{if } n \text{ is early or late} \\ 0 & \text{otherwise.} \end{cases}$$

5.1   Earlier Confidence Ratings Effects on End Confidence Rating

From Eq. 1, the confidence rating at the 10 s interval affected the end confidence rating by a factor of 0.21 ($p \leq 0.01$; importance = 0.15) and the confidence rating at the 20 s interval affected the end confidence rating by a factor of 0.24 ($p \leq 0.01$; importance = 0.16). As can be seen in Fig. 6, the confidence ratings during a run increase as the run continues. Thus, as the run continues, newer information influences the end confidence rating the most.

---

[1] The use of trademarks or names of manufacturers in this report is for accurate reporting and does not constitute an official endorsement, either expressed or implied, of such products or manufacturers by the National Aeronautics and Space Administration.

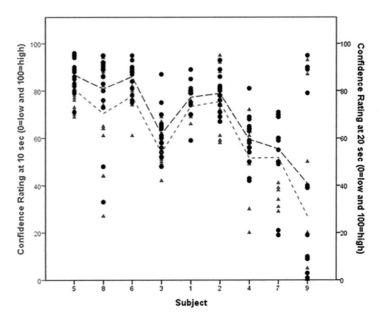

Fig. 6. Confidence rating at 10 s into run (gray triangles) and at 20 s into run (black circles) by subject. The gray dotted line indicates the average confidence rating at 10 s and the black dotted line indicates the average confidence rating at 20 s.

5.2   Subject Effects on End Confidence Rating

Subject personality had a significance of $p \leq 0.01$ and an importance of 0.30. There were 3 groups of subjects (Fig. 7). In general, subjects that had low frustration and good performance added 15 points to their end confidence rating and subjects that had some frustration and slightly lower performance with the task added 9 points to their end confidence rating. Subjects that had high frustration and poor performance on the task added no additional points to their end confidence rating (Fig. 8).

5.3   Scenario Level Effects on End Confidence Rating

Scenario level, or path deviation level, had a significance of $p \leq 0.01$ and an importance of 0.24 (Fig. 9). Scenarios which had a path deviation of at least ±14 in latitude, longitude, and altitude and ±9 in time decreased the end confidence rating by nine. Not surprisingly, vehicles that deviate quite a bit from the planned path resulted in a lower confidence rating.

5.4 Run Occurrence Effects on End Confidence Rating

Finally, when the run occurred had a significance of $p \leq 0.02$ and an importance of 0.10 (Fig. 9). The first and last runs, which added 6 points to the end confidence rating, were in a separate group from the middle runs. This may indicate an operator attentional lag in the middle of a mission.

5.5 Linear Regression End Confidence Rating Prediction

Figure 10 shows an example predicted end confidence rating for a subject in group 1 by run number. Absolute error was calculated using Eq. 2. The mean absolute error was 5.5 with an standard error of the mean of 1.6, maximum absolute error of 24.3 and a minimum absolute error of 0.2. Table 1 indicates the above values for all subjects. As can be seen in Table 1 and Fig. 10, the absolute mean error is approximately 10 indicating that the linear regression equation can predict the end confidence rating of a person fairly accurately.

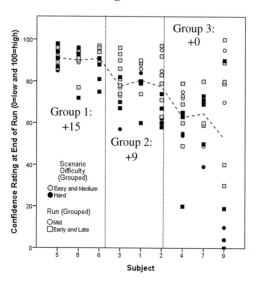

Fig. 7. End confidence rating by subject. The black dotted line indicates the average end confidence rating for each subject. Light grey indicates easy and medium scenario difficulty and black indicates hard scenario difficulty. Circles indicate mid-runs and squares indicate early and late runs.

$$Absolute\ Error = \left| \begin{array}{l} End\ Confidence\ Rating_{[Subject,\ Run]} - \\ Predicted\ End\ Confidence\ Rating_{[Subject_{Personality},\ Run_{Group},\ Path_{Deviation}]} \end{array} \right| \quad (2)$$

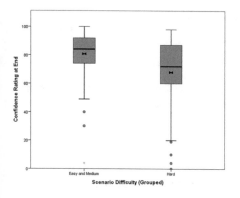

Fig. 8. End confidence rating by grouped scenario difficulty box plot. Bow tie indicates the mean.

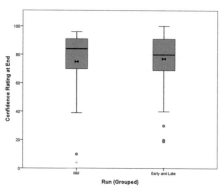

Fig. 9. End confidence rating by grouped run occurrence box plot. Bow tie indicates the mean.

Table 1. End confidence rating mean error statistics for each subject. Absolute error is calculated using Eq. 2.

Subject	Mean absolute error	Absolute standard error of the mean	Absolute maximum error	Absolute minimum error
5	5.5	1.6	24.3	0.2
8	9.5	2.8	40.4	0.5
6	6.0	0.9	13.8	1.5
3	9.8	1.8	23.0	1.2
1	7.0	1.4	17.4	0.1
2	8.7	1.5	20.7	0.0
4	8.3	1.5	21.3	0.2
7	10.4	1.8	22.3	0.3
9	24.6	3.6	43.1	0.1

6 Discussion

The above results indicate that an operator's confidence or trust can be predicted by objective measures (see Eq. 1). Each successive confidence rating builds on previous confidence ratings for a particular run. The function also is dependent on the subject's personality with regards to workload, how easily he may become frustrated, and his performance. The time or experience within a mission also affects trust. Lastly, if the vehicle has large deviations from the proscribed path, confidence decreases. Although the information provided on the display did not affect the linear regression, it did highlight the deviations; therefore, this variable may have rolled up into the scenario deviation variable. In general,

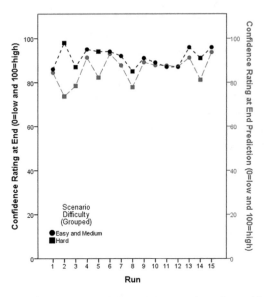

Fig. 10. Example end confidence rating of a subject (black icons) compared to predicted end confidence ratings (gray icons). Circles indicate easy and medium scenario difficulty. Squares indicate hard scenario difficulty.

$$Trust = f(personality, observed\ path\ deviation, experience).$$

7   Conclusions

A primary goal is for public and civil operators is to have one person managing several vehicles with different mission goals. To realize the full benefit from autonomy, these systems will have to react to unknown events and uncertain dynamic environments. The resulting number of behaviors is essentially infinite; thus, the system is effectively non-deterministic.

Even with the system effectively non-deterministic, an operator needs to understand and trust the actions of the autonomous vehicles in order for the system as a whole to operate optimally. This research began to tackle non-deterministic systems and trust by beginning to develop a user trust function based on intent information displayed and prescribed bounds on allowable behaviors/actions of the non-deterministic system. Linear regression shows promise on being able to predict a person's confidence of the machine's prediction. Linear regression techniques indicated that subject characteristics, scenario difficulty, the experience with the system, and confidence earlier in the scenario account for approximately 60% of the variation in confidence ratings.

However, these results are preliminary because this experiment entailed a few short runs with a limited subject pool. Additional and longer runs will better determine how time affects confidence. A larger subject pool will aid in determining more precise subject characteristics that affect confidence. This will simplify the trust function in that it will not have to be tuned to each individual. Furthermore, even though varying amounts of information were provided, the primary driving factors appears to be time and the amount of deviation from path. The information provided on the displays may have been rolled into the deviation from path variable. Therefore, additional research needs to be conducted in order to better refine the effects of deviation and the information provided.

In any case, trust appears to be a function personality, deviation, and time. With a parametric function, user trust could be predicted. With this prediction, additional information could be provided to maintain an appropriate level of trust. The appropriate level of trust maintained among team members, whether they be human or machine, will help enable a system to perform optimally. Lastly, providing detailed information regarding the reasoning behind the prediction (second level of the SAT model) may also beneficially affect the trust function.

Acknowledgments. This research was supported by NASA Langley Research Center IRAD funding in 2016.

References

1. Chua, L.O.: Chua circuit. Scholarpedia **2**, 1488 (2007)
2. Chua, L.O., Wu, C.W., Huang, A., Zhong, G.-Q.: A universal circuit for studying and generating chaos-Part I: routes to chaos. IEEE Trans. Circ. Syst. I Fundam. Theor. Appl. **40**, 13 (1993)
3. Beller, J., Heesen, M., Vollrath, M.: Improving the driver-automation interaction. Hum. Factors J. Hum. Factors Ergon. Soc. **55**, 11 (2013)

4. McGuirl, J.M., Sarter, N.B.: Supporting trust calibration and the effective use of decision aids by presenting dynamic system confidence information. Hum. Factors J. Hum. Factors Ergon. Soc. **48**, 10 (2006)
5. Verberne, F.M.F., Ham, J., Midden, C.J.H.: Trust in smart systems. Hum. Factors J. Hum. Factors Ergon. Soc. **54**, 11 (2012)
6. Couch, L.L., Jones, W.H.: Measuring levels of trust. J. Res. Pers. **31**, 18 (1997)
7. Jian, J.-Y., Bizantz, A.M., Drury, C.G.: Foundations for an empirically determined scale of trust in automated systems. Int. J. Cogn. Ergon. **4**, 16 (2000)
8. Hoff, K.A., Bashir, M.: Trust in automation: integrating empirical evidence on factors that influence trust. Hum. Factors J. Hum. Factors Ergon. Soc. **57**, 407–434 (2015)
9. Schaeffer, K.E.: The perception and measurement of human-robot trust. Doctor of Philosophy, p. 359, Department of Modeling and Simulation in the College of Sciences, University of Central Florida, Orlando, Florida (2013)
10. Boyce, M.W., Chen, J.Y.C., Selkowitz, A.R., Lakmani, S.G.: Effects of agent transparency on operator trust. In: Proceedings of the Tenth Annual ACM/IEEE International Conference on Human-Robot Interaction Extended Abstracts, pp. 179–180. ACM, New York (2015)
11. Chen, J.Y.C., Barnes, M.J., Selkowitz, A.R., Stowers, K.: Effects of agent transparency on human-autonomy teaming effectiveness. In: 2016 IEEE International Conference on Systems, Man, and Cybernetics (SMC), pp. 1838–1843. IEEE (2016)
12. Chen, J.Y.C., Procci, K., Boyce, M.W., Wright, J., Garcia, A., Barnes, M.J.: Situation awareness-based agent transparency, p. 36. Laboratory, U.S.A.R, U.S. Army Research Laboratory, Aberdeen Proving Ground (2014)
13. Lakhmani, S., Abich, J., Barber, D., Chen, J.: A proposed approach for determining the influence of multimodal robot-of-human transparency information on human-agent teams. In: Schmorrow, D.D., Fidopiastis, C.M. (eds.) Foundations of Augmented Cognition: Neuroergonomics and Operational Neuroscience: 10th International Conference, AC 2016, Held as Part of HCI International 2016, Toronto, ON, Canada, 17–22 July 2016, Proceedings, Part II, pp. 296–307. Springer International Publishing, Cham (2016)
14. Mercado, J.E., Rupp, M.A., Chen, J.Y.C., Barnes, M.J., Procci, K.: Intelligent agent transparency in human-agent teaming for multi-UxV management. Hum. Factors **58**, 401–415 (2016)
15. Wright, J.L., Chen, J.Y.C., Barnes, M.J., Hancock, P.A.: Agent reasoning transparency's effect on operator workload. Proc. Hum. Factors Ergon. Soc. Annu. Meet. **60**, 249–253 (2016)
16. Endsley, M.R.: Toward a theory of situation awareness in dynamic systems. Hum. Factors **37**, 32–64 (1995)
17. Endsley, M.R.: Situation awareness misconceptions and misunderstandings. J. Cogn. Eng. Decis. Making **9**, 4–32 (2015)
18. Parasuraman, R., Riley, V.: Humans and automation: use, misuse, disuse, abuse. Hum. Factors J. Hum. Factors Ergon. Soc. **39**, 230–253 (1997)
19. Moray, N., Inagaki, T., Makoto, I.: Adaptive automation, trust, and self-confidence in fault management of time-critical tasks. J. Exp. Psychol. Appl. **6**, 44–58 (2000)
20. http://www.chuacircuits.com/
21. Trujillo, A.C.: How electronic questionnaire formats affect scaled responses. In: 15th International Symposium on Aviation Psychology, Dayton, OH (2009)

Traditional Vs Gesture Based UAV Control

Brian Sanders[1]([X]), Dennis Vincenzi[2], Sam Holley[2],
and Yuzhong Shen[3]

[1] Department of Engineering and Technology, Embry-Riddle Aeronautical
University, Worldwide, Daytona Beach, USA
sanderb7@erau.edu
[2] Department of Aeronautics, Graduate Studies, Embry-Riddle Aeronautical
University, Worldwide, Daytona Beach, USA
{vincenzd, holle710}@erau.edu
[3] Department of Modeling, Simulation, and Visualization Engineering,
Old Dominion University, Norfolk, VA, USA
YShen@odu.edu

Abstract. The purpose of this investigation was to assess user preferences for controlling an autonomous system. A comparison using a virtual environment (VE) was made between a joystick based, game controller and a gesture-based system using the leap motion controller. Command functions included basic flight maneuvers and switching between the operator and drone view. Comparisons were made between the control approaches using a representative quadcopter drone. The VE was designed to minimize the cognitive loading and focus on the flight control. It is a physics-based flight simulator built in Unity3D. Participants first spend time familiarizing themselves with the basic controls and vehicle response to command inputs. They then engaged in search missions. Data was gathered on time spent performing tasks, and post test interviews were conducted to uncover user preferences. Results indicate that while the gesture-based system has some benefits the joystick control is still preferred.

Keywords: Autonomous systems · Leap Motion Controller
Gesture-based interface

1 Introduction

The autonomous systems market continues to grow with systems (aerial and ground) ranging from those that can be held in the palm of your hand to larger, extremely capable military systems. For recreational use, and an emerging business market, the FAA requires an unenhanced line of sight to the vehicle be maintained. However, new technology, in the form of mixed reality headsets are emerging and accompanied by a plethora of gesture-based systems (see for examples, LMC, Oculus with Touch) that may enable a future change in policy and regulations. These head mounted display systems combined with gesture-based command control interfaces offer a new approach to vehicle control that is unencumbered by traditional hand held devices, and offer unique capabilities for mission planning and vehicle, even multivehicle, control.

© Springer International Publishing AG, part of Springer Nature 2019
J. Chen (Ed.): AHFE 2018, AISC 784, pp. 15–23, 2019.
https://doi.org/10.1007/978-3-319-94346-6_2

The design of these systems will require careful investigation of human factors issues to populate gesture libraries that are natural and intuitive, as well as cognitive loading considerations due to the easy availability of a vast amount of visual information. This combination of gesture libraries and cognitive loading is the focus of this research.

1.1   New Technology for Displays and Controls

The concept of a reality-virtuality continuum was first introduced by Paul Milgram in 1994. In that paper, Milgram and Kishino [1] discuss the concept of a reality-virtuality continuum with the real-world environment on one end of the continuum and a totally virtual computer-generated environment on the other end of the continuum. Between the two ends of the continuum is a wide range of mixed reality variations between total real environment and total virtual environment. Most advanced interfaces today fall somewhere in the mixed reality section of the reality-virtuality continuum.

State of the art technology began revolutionizing the Human-Machine Interface with the movement away from legacy control devices such as joysticks and other physical controls, toward more innovative interface technologies such as touchscreens, virtual reality displays (VR), augmented reality displays (AR), and mixed reality displays (MR). VR displays or VR-like displays are now affordable and commonplace, and are regularly used as the display of choice when immersion into a 3D environment is preferred. See-through displays offer the ability to create AR or MR displays by allowing the real world to be viewed through a see-through display that can be used to augment the real world with additional information. High resolution displays on phones or inexpensive Helmet Mounted Displays (HMDs) such as the Oculus Rift have begun to replace and augment visual systems to develop environments that serve as displays for vehicle parameters as well as provide an egocentric view from the UAS camera [2, 3].

The creation of a MR display that integrates both the real world and computer-generated world into a display that uses relevant parts of both environments is meant to create a display that is more effective than either one by itself. The goal of this MR type display would be to integrate relevant portions of both the real world and virtual world to produce a display that is efficient, functional, user friendly, and (hopefully) intuitive in design. The capability exists for technology to provide more information with realistic visual perspectives similar to looking through a Heads-Up Display (HUD) on a manned aircraft. Utilization of these types of technologies, if designed correctly, can result in a more realistic visual display that provides the information needed for successful operation with minimal training requirements [2].

Although VR and MR type displays have been available since the 1980s and 1990s, MR interfaces that include control components have not. Interactive, MR display and control interfaces have only recently appeared on the consumer market in a usable and affordable form. Typically, a combination of technology can be integrated and utilized to create an inclusive human-machine interface that can be used to both display information in a VR environment while designing a control interface which can be used to manipulate objects in the real or virtual worlds. This combination of technology provides the means to design a MR display and a VR control interface for use in a real or virtual world. Inexpensive hardware used in this experiment include the Leap Motion

Controller (LMC) and Oculus Rift Goggles. The LMC generates the operational gesture recognition environment while the Oculus Rift Goggles provides an immersive or see-through information display environment.

1.2   Cognitive Loading

During the early stages of evolution for gesture-based control applications, cognitive processing and loading issues [4] are introduced as factors that will be incorporated gradually as the study advances. For the present effort, conditions aligned with aerodynamic factors and perception in contrasting environments will be observed. Initially, attention is directed to features of direct viewing in natural settings compared with views from a tablet displaying virtual images and information. This is particularly apparent when switching views between operator real-world view and a virtual framework. Recent evidence indicates that very different brain processes are involved in comprehending meaning from these sources [5], in particular competition among select hippocampal neurons, suggesting accelerated depletion of neural resources under high task loadings [6]. As stated earlier, anecdotal observations will accompany the tasks performed in the current trials to identify particular variables for study later.

As elements and workload increase in number and complexity, task loading will follow. The influences of channelized attention, sensory cues, and loss of energy state awareness [7] are potential targets for further study. In turn, influences on cognitive mapping [8] will be investigated to elaborate effects on sustained attention, conflict resolution, and rapid updating of working memory [9]. Another element for later focus in the current line of research is to evaluate perceptual and cognitive processing features when using a three-dimensional virtual image with an environment that entails moving units or elements. As might be expected, cognitive loading considerations will likely increase [10] in presence and influence. In the intermediate stages of this research trajectory, when operators will use a heads-up display, combined effects of real-world and virtual cognitive processing will be an intense area of inquiry, including consequences of refractory periods and protein cycling limits in memory buffering [11] as they affect vigilance and judgment.

The above discussion highlights some of the opportunities and considerations for the control of autonomous vehicles using head mounted and gesture-based systems. Previous work by the authors involved the development of a gesture-based library for control of a drone [2]. In that paper, 11 basic control functions were identified along with the associated hand gestures to achieve the desired response. This work builds upon that by examining user's ability to control a representative drone in a virtual environment (VE).

2   Virtual Environment and Controllers

The control approaches were compared using a virtual environment (VE) created in Unity3D (™). A partial shot of the VE is shown in Fig. 1 below. Its overall purpose was to enable a scenario that included some depth and scale perception. The drone is a generic representation of a recreational quadcopter drone. It models a 1 kg drone with nominal dimensions of 30 cm × 30 cm × 10 cm and has red lights indicating the forward part of the drone and blinking green lights in the rear of the unit. The VE is of a generic rolling hills environment that contains some natural environment features such as trees, a lake, and several man-made features such as a jeep, tents, and recreational vehicles. The latter were used in the second part of the test when the participant was asked to locate and proceed to one of these targets.

Fig. 1. Virtual environment (probably need one with the Jeep in it or some other target)

The screen contained features to inform the participant about the status of the vehicle. The red letters are a Heads Up Display that provides information on the altitude, speed, and distance to the vehicle. At this point the test is done in screen space mode, but as the research continues and VR and Mixed Reality Goggles are incorporated into the configuration, these features should transition nicely. There is also a directional arrow in the left-hand corner that indicates the direction of the vehicle. It points in the forward direction (i.e., red lights). The green box is an indicator for when the vehicle responds to the virtual hands. Outside of this area other commands can be given, such as turning the vehicle camera on and off.

The commands are given by either a traditional joystick-based controller, or gestures captured by a leap motion controller. The joystick is shown in Fig. 2 and is a typical Xbox 360 controller. The left stick controls the rotation of the vehicle and the fore/aft translation (Fore/aft translation referring to direction of the lights). The right stick controls the altitude and the left/right translation. Finally, the A button changes the perspective from the operator to a view from the drone camera and the B button resets the vehicle to the starting position.

The gesture commands were captured using a leap motion controller (LMC). Figure 3 shows the leap motion sensor and its coordinate system. It is a right-handed coordinate system, as opposed to Unity3D which uses a left-handed coordinate system. The LMC is able to capture hand motions with a sub-millimeter accuracy (Weichert et al.) [12]).

Rotation

Reset UAV Position

Change Camera View

Altitude

Fore/Aft
Translation

Lateral Translation

Fig. 2. Joystick controller

Multiple hand models come with the LMC. These range from basic hand models shown in Fig. 4 below to robotic looking hands to humanoid looking hands. For this investigation the user could roll the left hand about the z-axis for the vehicle to yaw and pitch it about the x-axis for a forward and aft motion. A rotation of the right hand around the z-axis results in a left/right translation of the vehicle while a right hand rotation around the x-axis results in a change in altitude. One final capability was via the use of the LMC based user interface to control the camera. It is a slick capability available in the Orion Version of the API [13]. In this case a UI is attached to the left hand and is visible when that hand is rotated toward the user as shown in Fig. 5 below. In this case the interface enabled the view selection from either the view of the operator or the view from the camera. The view from the camera assists in identifying targets.

+Y

+X

+Z

Fig. 3. Leap motion controller and coordinate system [13]

3   Methodology

The intent of this phase of the research was to determine a participant's control approach preference and what led to that decision. To help the user make an informed decision, each participant was exposed to a play environment and an operational environment. No tasks were assigned to the user in the play environment. The play environment was meant to allow the participant to become familiar with the basic controls and sensitivity of the vehicle while using either the joystick or gesture-based command inputs. While it did contain several of the elements to help with scale and

Left Hand
Pitch: Fore/Aft Translation
Roll: Rotation

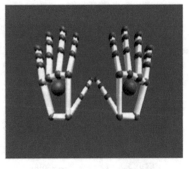

Right Hand
Pitch: Altitude
Roll: Lateral Translation

Fig. 4. Leap motion UI for controlling camera view

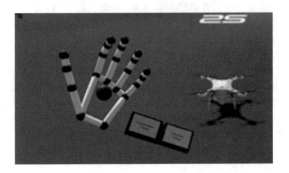

Fig. 5. Leap motion UI for controlling camera view

depth perception, it did not contain any of the targets mentioned above. In the operational environment users were tasked to find a target (i.e., ground of tents), proceed to that target, and then land in an identified landing zone. The landing zone had a 10 m radius, so it was quite large compared to the drone.

This approach provided some insight into how well a user could perform a task without being overly burdensome by asking them to fly specific flight plans. The flying of precision flight paths will be reserved for future investigations where assessment of accuracy and performance are needed. The qualitative data gathering from this was the amount of time spent in play and the amount of time to complete the assigned task in the operational environment. Post test interviews consisted of a series of six questions measured on a Likert scale and constructed to uncover the perceived suitability and advantages of the gesture-based control approach, as well as other features the simulation should include to enhance usability and functionality of the interface.

4   Results and Discussion

Two sets of tests were conducted to date. The first set of tests included two participants that were recreational drone pilots. The purpose was to provide feedback on the basic virtual environment requirements. For this early work it was desirable to keep the cognitive loading to a minimum, so the operator could focus on the control system, and the feedback indicated the fundamental data related to vehicle orientation and scaling references that need to be included. The two participants help guide what basic information needed to be included which was: textural information related to altitude, speed, and slant range between the operator and the drone. They also suggested that some objects be placed in the scene to help process scaling issues related to distance. The final scene represented a camping area at a lake with recreational vehicles at one location and a group of tents at the other.

The second set of tests included four participants. The purpose of these tests was to provide feedback on the two control system approaches. As mentioned above each participant engaged in four scenarios. Two in play (one with the Joystick and one with the LMC system) and similarly two in a search mission, such as looking for and traveling to the RV park. On average twice as much time (11 min vs 22 min) was spent in play mode with the gesture-based system. Users were able to quickly feel comfortable with the joystick approach. On the other hand, while the controllability significantly improved from the start the users still did not feel as comfortable at the end of the play session with LMC system as compared to the joystick. Likewise, mission times for the joystick were on the order of 3 min while the missions for the LMC were rarely completed due to fatigue and frustration with the system. This suggests that a change in test approach may be required. For example, as opposed to running a participant through all four scenarios maybe just have them focus on one or the other control system.

Total test time and post interview per participant took just under an hour, and the participants reported feeling fatigued at the end, mostly due to using the LMC system. It is thought that this is due to the hand gestures requiring more energy compared to the finger motion that can be used with the joystick. Even though it was more energy the users like that it made them feel more connected to the vehicle response. Also, use of hand gestures is a newer approach so the cognitive load was most likely higher since they were processing more (i.e., the visuals of the hands) visual information and correlating the vehicle response to their inputs. The users also commented that they preferred the joystick for making small command inputs. This notion is consistent with what is reported by Weichert et al. [12] when they compared the accuracy of the mouse to that of the LMC. The LMC is highly accurate when it comes to detecting the hand placement and motion, but the challenge is transforming that information into precise control such as the vehicle in this case. Achieving the same control precision with hand motions may require some additional filtering of the hand movements translated to command actions. The addition of filtering algorithms, automation, or some form of artificial intelligence to dampen, regulate, adjust, or interpret gestures intended to control a vehicle is something to be considered for future iterations.

For the most part the visual content was satisfactory for the participants. The location of the textural information was enough, and the users' responses did not indicate they were overly taxing of information processing. In fact, they were typically so focused on the vehicle that they needed to be told this information was available. On the other hand, the virtual hands were distracting. This concern was alleviated by making them smaller. It still provided a point of reference, but it could be accessed when needed rather than constantly in the visual processing path. Thus, the larger hands were a distraction and loaded up the visual processing apparatus. While the smaller hands still provided an orientation but did not seem to distract attention.

5 Summary and Conclusions

The promise of emerging technology for the design of new control environments for autonomous systems is alluring. These new systems will enable a departure from using highly accurate (i.e., joystick), low information visual systems to less accurate but more flexible gesture-based systems accompanied by a visual system that can rapidly provide vast amounts of information. In the present case, the LMC itself is highly accurate, but this particular application design has not transferred that accuracy to a system control approach yet.

References

1. Milgram, P., Kishino, F.A.: Taxonomy of mixed reality visual displays. IEICE Trans. Inf. Syst. **E77-D**(12), 1321–1329 (1994)
2. Sanders, B., Vincenzi, D., Shen, Y.: Gesture based UAV control. In: Proceeding from the 8th International Conference on Applied Human Factors and Ergonomics, Los Angeles, CA, 17–21 July 2017
3. Chandarana, M., Meszaros, E., Trujillo, A., Allen, B.: Challenges of using Mulitmodal HMI for unmanned mission planning. In: Proceeding from the 8th International Conference on Applied Human Factors and Ergonomics, Los Angeles, CA, 17–21 July 2017
4. Antonenko, P., Paas, F., Grabner, R., van Gog, T.: Using electroencephalography to measure cognitive load. Ed. Psychol. Rev. **22**(4), 425–438 (2010)
5. Ravassard, P., Kees, A., Willers, B., Ho, D., Aharoni, D., Cushman, J., Aghajan, Z., Mehta, M.: Multisensory control of hippocampal spatiotemporal selectivity. Science **340**(6138), 1342–1346 (2013)
6. Ungar, M.: The social ecology of resilience: addressing contextual and cultural ambiguity of a nascent construct. Am. J. Orthopsychiatr. **81**(1), 1–17 (2011)
7. Dodd, S., Lancaster, J., Miranda, A., Grothe, S., DeMers, B., Rogers, B.: Touch screens on the flight deck: the impact of touch target size, spacing, touch technology and turbulence on pilot performance. In: Proceedings of 58th Annual Meeting of the Human Factors and Ergonomics Society, pp. 6–10, Chicago (2014)
8. Gevins, A., Smith, M.E., Leong, H., McEvoy, L., Whitfield, S., Du, R., Rush, G.: Monitoring working memory load during computer-based tasks with eeg pattern recognition methods. Hum. Factors **40**(1), 79–91 (1998)
9. Bowers, M., Christensen, J., Eggemeier, F.: The effects of workload transitions in a multitasking environment. In: Proceedings of 58th Annual Meeting of the Human Factors and Ergonomics Society, pp. 220–228, Chicago (2014)

10. Rorie, R., Fern, L.: UAS measured response: the effect of GCS control mode interfaces on pilot ability to comply with ATC slearances. In: Proceedings of 58th Annual Meeting of the Human Factors and Ergonomics Society, pp. 64–68, Chicago (2014)
11. Cottrell, J., Levenson, J., Kim, S., Gibson, H., Richardson, K., Sivula, M., Li, B., Ashford, C., Heindl, K., Babcock, R., Rose, D., Hempel, C., Wiig, K., Laeng, P., Levin, M., Ryan, T., Gerber, D.: Working memory impairment in calcineurin knock-out mice is associated with alterations in synaptic vesicle cycling and disruption of high-frequency synaptic and network activity in prefrontal cortex. J. Neurosci. **33**(27), 10938–10949 (2013)
12. Weichert, F., Bachmann, D., Rudak, B., Fisseler, D.: Analysis of the accuracy and robustness of the leap motion controller. Sensors **13**(5), 6380–6393 (2013). https://doi.org/10.3390/s130506380
13. Leap motion developer: https://developer.leapmotion.com/. Accessed 2 March 2017

Establishing a Variable Automation Paradigm for UAV-Based Reconnaissance in Manned-Unmanned Teaming Missions

Experimental Evaluation and Results

Christian Ruf[✉] and Peter Stütz[✉]

Institute of Flight Systems (IFS), University of the Bundeswehr Munich
(UniBwM), Neubiberg, Germany
{christian.ruf, peter.stuetz}@unibw.de

Abstract. This work addresses the factor of degraded automation reliability of machine based aerial reconnaissance in a manned-unmanned teaming approach. An army transport helicopter is accompanied by three unmanned aerial vehicles for reconnaissance purposes, guided by the helicopters crew. Automated capabilities onboard the UAVs offer high automated, task-based guidance as well as manual operation. We designed and implemented an assistance system in our helicopter flight simulator, that supports the commander in gaining relevant reconnaissance information on flight routes for the helicopter to follow. Due to imperfection in automated reconnaissance performed by machine algorithms, we explicitly regarded the aspect of degrading reliability by utilizing the paradigm of "Levels of Automation". The automation system produces reconnaissance results, thereby considering differing automation reliability. Several data representation modes were applied to display preprocessed results in the helicopters multi-function displays. We conducted an extensive human-in-the-loop campaign with army helicopter crews in full mission scenarios, in which system-triggered changes between the automation levels occurred and the cooperative human-machine relationship changed online. This paper presents questionnaire-gathered results of our investigation during mission execution, shedding light on human factors, user acceptance and system design aspects.

Keywords: MUM-T · Multi-UAV · Levels of automation
Adaptive automation · Automation reliability · Automation trustworthiness
Assistance system · Human factors · Mental workload · Trust in automation
Human-in-the-loop experiment

1 Introduction

One of the research fields at the Institute of Flight Systems (IFS) of the University of the German Armed Forces (UniBwM) is the design, prototypical implementation and experimental evaluation of operation support and assistance functionalities for mission sensors (reconnaissance sensors) carried by unmanned aerial vehicles (UAVs). In our application, three UAVs (multi-UAV) are guided by a transport helicopters (HC) crew

© Springer International Publishing AG, part of Springer Nature 2019
J. Chen (Ed.): AHFE 2018, AISC 784, pp. 24–35, 2019.
https://doi.org/10.1007/978-3-319-94346-6_3

and form a manned-unmanned team (MUM-T) with the two-seated helicopter itself to perform transport missions in our helicopter mission simulator (Fig. 1).

Fig. 1. MUM-T configuration in our helicopter mission simulator

The purposive deployment of automated UAVs for reconnaissance of HC flight routes and mission places in full mission scenarios brings in new and impactful human factors aspects for the helicopters commander, compared to the regular task spectrum of a regular transport helicopter flight.

2 Background and Method

2.1 Problem Scope

One the one hand, the commander's necessary work-processes to handle the additional reconnaissance instruments and to derive and include relevant information in the mission progress are expected to induce additional mental workload (MWL) during system operation. We use a representation for MWL described in [1], with a task-model containing modelled human resources customized to our interfaces, interaction modalities and display objects of visualized reconnaissance results [2]. On the other hand, the automation of airborne reconnaissance brings in domain-specific conditions. Algorithm based sensor data evaluation and automatic target recognition (ATR) systems often do not perform in a perfect or highly deterministic way, e.g. because of imperfection or typical artefacts in sensor data evaluation or varying operation environments [3]. Out of a technical perspective, this circumstance can be addressed by the measure of "trustworthiness", describing the reliability of automated systems. In [4], a performance prediction method for image assessment algorithms is proposed. An operator's confidence in automated systems is considered as "Trust in automation" [5].

2.2   Method

We addressed the two human factors of the crew's MWL-situation and "Trust in automation" in a human-centered [6], adaptive [7] automation approach by utilizing the paradigm of *"Levels of Automation"* [8, 9]. Therefore, we designed and implemented an assistance system (basic ideas described in [3]) that supports the helicopters commander on airborne reconnaissance for automated search of HC flight paths that are free of threats as well as manual operation of reconnaissance capabilities for self-protection. The assistance system offers variable automated functions from the domain of machine based data processing, especially algorithm based data assessment, data transformation and storage. We also created an automated result management, different methods for result representation and functionality for dynamic user involvement applying interface automation. The interface automation adapts the user interface dynamically to a task-centered layout configuration. This work flow optimization should relax upcoming workload peaks.

The technical functions are separated in three levels of automation [2, 3] that can be alternated online during mission execution. The crew realizes these automation modes by different types of data representation and user involvement strategies. The basic idea is that a lower automation degree involves more crew resources [8, 10], but also achieves beneficial reconnaissance performance due to the humans contribution. The methods principle is to change the automation levels. If the reliability of the algorithm performance decreases, the automation level is changed to a lower one where less preprocessing is applied. If an excessive workload situation is predicted in [11] by applying model-based resource determination [1], a task scheduler involves the commander by utilizing the interface automation to present urgent targets to identify. Triggers for changing the automation level are the crew's MWL-situation (determined online in [1, 11]) as well as the reconnaissance performance (result trustworthiness).

In detail, the automation modes are:

- *"Annotated Video"*: Live video feed with annotation of detected object candidates available in the HC-cockpit.
- *"Map Assisted"*: Visualization of reconnoitered routes on a tactical map by colors and shading of merged sensor footprints during automated reconnaissance. This mode offers also rectified and georeferenced ground images (image mosaicking) on higher zoom levels of the map. Detections with low trustworthiness ("Target Candidates") are presented by an own designed, salient symbol (Fig. 2, right). Those have to be identified by the commander.
- *"Tactical Assisted"*: Visualization of reconnoitered routes, with detected and identified objects represented by tactical common warfighting symbolism [12] (Fig. 2, left) on a high level of automation trustworthiness. If "Target Candidates" appear, the user gets involved by display automation to identify those that appeared on the active helicopter route (Fig. 3).

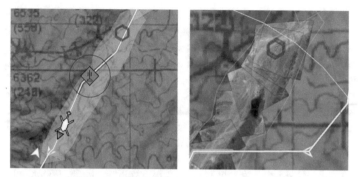

Fig. 2. Reconnaissance results on tactical map. Left: HC routes with common warfighting symbolism and "Target Candidates". Right: ground images and "Target Candidates"

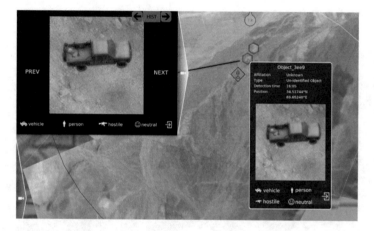

Fig. 3. Work flow optimization by display automation (Pop-up of top-left window as identification help) in "Tactical Assisted"-Mode; right window (retrievable at any time) shows still-image of detected and captured "Target Candidate" with identification input bar

3   Application and Experimental Evaluation

We conducted an extensive human-in-the-loop experimental campaign with four military transport helicopter crews of the *German Bundeswehr* in our helicopter flight and mission simulator (Fig. 1). Focus of the campaign was to observe the crew's usage of the UAV reconnaissance capabilities, their interactions with the variable automated reconnaissance system and the reconnaissance results.

3.1   Experimental Design

We started with a training phase of ~two days, in which the crews trained the employment of the UAVs from the cockpit to gain reconnaissance results from the automation system, learned to interpret and make usage of reconnaissance results and get used to the cockpit automation. The tutorials comprised detailed instructions on handling and functional background knowledge of the functionalities. Each function group was initially introduced and trained in an isolated scenario, later all function-alities were combined applied in two training missions that were accompanied by instructors, if necessary.

Afterwards, each crew performed five complete MUM-T transport missions (35–60 min) within three days on their own. The mission goals were to execute different transport tasks (troops, casualties, cargo) in different mission types (material transport, CASEVAC, MEDEVAC). There were several varieties within the missions, some contained mission updates, follow-up-missions or time-critical events. Each mission started with a detailed briefing that was held in a military style, afterwards the crews entered the cockpit and executed the mission. Thereby, three UAVs had to be guided by a high-level task-based-guidance paradigm [13] to find HC flight paths that are free of threats in a closed-loop-operation. Figure 4 shows an example of such a mission in a digital map plot.

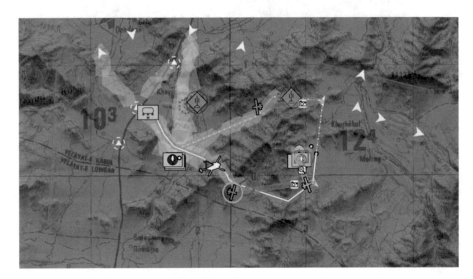

Fig. 4. Full-mission scenario with three UAVs for automated HC flight path reconnaissance

In this dynamic missions, the automated reconnaissance detection reliability and for this reason the automation modes and data representation modes changed automati-cally. Threats (armed ground vehicles) and "Target Candidates" to identify (hostile and civil ground vehicles) appeared dynamically and were presented online on the heli-copters MFDs (multi-functional displays) during mission execution.

3.2 Participants

The four helicopter commanders (average age: 51,5 years) have on average 1950 h of experience as Pilot in Command (PIC), on average 300 h in military operations.

4   Results

In this paper, we focus on a human factors-centered question. Those human factors aspects were gathered by questionnaires. We evaluated MWL afterwards by NASA-TLX questionnaires. Furthermore, we gathered individualized human factors aspects by individualized questionnaires at the campaigns end.

After the missions, we asked about the degree of realism of the simulation environment (Fig. 5) to estimate the degree of immersion into the simulated environment.

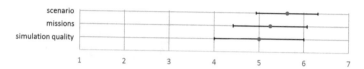

Fig. 5. Degree of realism of simulation environment

The commanders rated the simulation as quite realistic in terms of scenario, missions and simulation quality.

4.1   Human Factors: Mental Workload

Figure 6 shows the perceived workload by NASA-TLX for all five missions.

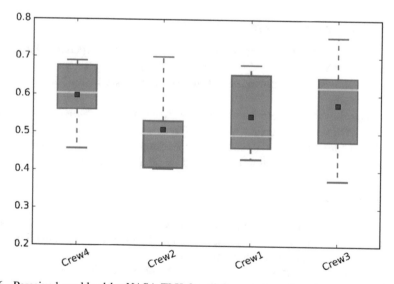

Fig. 6. Perceived workload by NASA-TLX for all five missions (boxplot representation with small red square representing mean)

We observed correspondences between the commanders' work flow, behavior, operator errors during the experiment and the load index values. The ratings seem to be similar, but there is one anomaly (crew 2). Further analysis has to be done here.

The commanders rated the demands on effort, concentration and attention on task execution as shown in Figs. 7 and 8. The term "task" comprises all activities concerning handling and supervising the reconnaissance system and the produced reconnaissance results on the MFDs. Green cells mark a positive result, red ones a critical.

	fully disagree		neutral		fully agree
Effort					
Task completion was classified as demanding	•	••	•		
Task completion was experienced as easy			•	••	•
Concentration					
Appropriate task completion required concentration	•		•	••	
Task completion required continous concentration	•		•	••	
Memorizing task-relevant information required a high level of concentration	•	•	•	•	

Fig. 7. Commanders' rating on effort and concentration during task execution

The commanders' ratings concerning the demand on effort and attention were good, the rating on concentration demand was mixed.

4.2 Human Factors: Trust in Automation

Regarding an operator's trust in automation, we asked for their experienced global trust in automated reconnaissance and their trust in different data formats.

Figure 9 shows the commanders' specification of trust in the simulated reconnaissance system.

We asked the commanders whether they were aware of different credibility of the data formats. Only two commanders had different trust in the correctness of the data representation formats. Figure 10 shows their grading of the data formats produced by the automation system depending on different automation reliability.

4.3 Ergonomics

We asked the commanders to rate the perceptibility and clarity of the visualized reconnaissance results. They agreed in general that the automation system offered well-designed data representation methods (Fig. 11).

As well, positive ratings for the design and saliency of the "Target Candidates" were given (Fig. 12).

In the conclusion, we use this data to reveal a possible correlation between distrust in automation and misunderstanding of data representations.

Attention	fully disagree		neutral		fully agree
Task completion was classified as requiering active attention		●		●●●	
A low level of attention was experienced on task completion	●	●●●			
Recognition and extraction of task-relevant information was experienced as easy				●●●	●
Memorizing task-relevant information was experienced as easy			●	●●●	
Shifting attention to reconnaissance tasks was experienced as easy				●●	●●

Fig. 8. Commanders' rating on attention during task execution

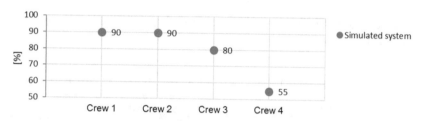

Fig. 9. Global trust [%] in automated simulated reconnaissance

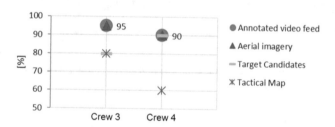

Fig. 10. Trust [%] related to different data display formats

Different types of data representation	fully disagree		neutral		fully agree
Meaning of used data representation methods was clear				●●	●●
Data representation methods required habituation process	●	●	●	●	
Different forms of information representation required effort to work with	●●	●●			
Different forms of data representation required effort to switch between		●●●●			

Fig. 11. Rating of user acceptance of different types of data representation

4.4 Automation Transparency

During mission execution, automation states changed and events of occurrence of detected objects appeared. We asked the commanders (Fig. 13) if they were aware of

	fully disagree		neutral		fully agree
Saliency of "Target Candidates"					
Symbolic represetation stands out of map smybology				•	•••
Shape and color are clearly indetifiable				••	••
Symbol is easy to perceive visually				•	•••
Symbol is catchy and memorable				••	••
Popping-up of new symbols on the map attracts visual attention quickly				•••	•

Fig. 12. Rating of user acceptance of symbolism of "Target Candidates"

the automation state and noticed the appearance of new detections that needed to be identified ("Target Candidates").

	fully disagree		neutral		fully agree
Automation Transparency					
Target Candidates appreared if reconnaissance capabilities degraded				•	•••
New Target Candidates appeared			•	••	•
Human input was needed for classification of Target Candidates				•••	•

Fig. 13. Automation transparency: Rating of understanding of automation state and outputs

These ratings shall be used to reveal a possible dependency of trust in automation on automation transparency.

4.5 Cooperative Human-Machine-Relationship

During the missions, the commander had the full authority of the automation system. We offered the ability to execute reconnaissance tasks in a fully automated fashion (including flight guidance and sensor operation). However, also an ad-hoc take-over and manual guidance of the reconnaissance sensors onboard the UAVs was possible. During automated task performance, commander and automation acted in a supervisory control relationship. We asked the commanders about the fulfillment of their requirements on authority and the experienced flexibility and adaptability of the assistance system to their own needs (Fig. 14).

In general, the ratings for user authority and the flexibility to own needs were good or rather very good.

4.6 Required Performance

We asked for the required performance of automation functions in an operational system. Three commanders estimated a minimum required performance level of such automation functions (Fig. 15).

	fully disagree		neutral		fully agree
Authority					
Complete fulfillment of users' control authority and controllability of automation			●	●●	●
Ability of user intervention, take-over and manual guidance of sensors is useful			●	●	●●
Flexibility/ Adabtibility to own needs					
Full adaptibility to users' spontaneous requirements in unforeseen cases					●●●●
Ability of fully manual guidance and complete user authority is essential					●●●●
Manual sensor guidance is an essential complement to automated sensor operation					●●●●
Experienced manual sensor guidance completes automated operation usefully					●●●●

Fig. 14. Rating of the cooperative relationship between human and automation

Fig. 15. Minimum required performance level of an operational system

The commanders also estimated the time-dependent usability of reconnaissance data as shown in Fig. 16. The questionnaire required three classes of currentness of data to specify for this reconnaissance setup.

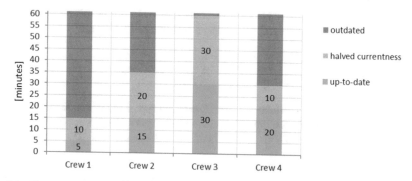

Fig. 16. Time dependent usability of reconnaissance results (numbers are durations in minutes)

This value can be used in future to parametrize the dynamic map representation of reconnaissance data (flight routes) on the MFDs, whereat outdated data is labelled or removed from the map. In the experiment, reconnaissance data of flight paths that were older than 15 min were removed. As well, this value can be used in future to specify the time gap and distance between the UAVs that gather data and the helicopter.

5   Conclusion and Discussion

We implemented an assistance system offering variable automation for UAV reconnaissance. We explicitly addressed the technical shortcoming of degrading reconnaissance capability by utilizing several levels of automation and data representation methods. We evaluated user acceptance, user requirements fulfillment and system specifications in an extensive human-in-the-loop campaign. In the results chapter, we built up a chain of causalities for evaluation, beginning from a simulation environment that was experienced more than satisfactorily. As well, the ratings of workload (NASA-TLX) were similar for the groups. This permits to concentrate on the evaluation aspects presented in the results chapter. The human factor of demanded concentration during task performance was mixed rated; a reason for that could be that the approach of target identification differed among the test candidates. Some assessed the targets directly after detection, others did a "bunch classification" after a series of detections. We assume a correlation between this and the concentration markings.

The demands on effort and attention were rated as good. Regarding the ratings of trust, only half of the crews had different trust in different data representations. Reasons for that may either be still present missing awareness of the systems behavior or the fact that their rating related on the absence or presence of the data itself, not their pre-processing stage.

The rating of trust in automation and trust in data representations seem to be plausible, because we asked for the perceptibility and clarity of the data themselves to avoid that trust depends on information misunderstanding. As well, missing transparency of the automation system can be excluded as influencing the ratings on trust levels. The assistance systems flexibility, the offered adaptability to own needs as well as the concept of user authority met with great approval. The proposed method based on levels of automation received a high user acceptance in general.

References

1. Honecker, F., Brand, Y., Schulte, A.: A task-centered approach for workload-adaptive pilot associate systems. In: Proceedings of the 32nd Conference of the European Association for Aviation Psychology – Thinking High and Low: Cognition and Decision Making in Aviation (2016)
2. Ruf, C., Stütz, P.: Model-driven payload sensor operation assistance for a transport helicopter crew in manned-unmanned teaming missions: assistance realization, modelling and experimental evaluation of mental workload. In: Engineering Psychology and Cognitive Ergonomics: Performance, Emotion and Situation Awareness. EPCE 2017. LNCS, vol. 10275. Springer, Cham (2017)

3. Ruf, C., Stütz, P.: Model-driven sensor operation assistance for a transport helicopter crew in manned-unmanned teaming missions : selecting the automation level by machine decision-making. In: Advances in Human Factors in Robots and Unmanned Systems, vol. 499, pp. 253–265 (2016)
4. Hellert, C., Stütz, P.: Performance prediction and selection of aerial perception functions during UAV missions. In: IEEE Aerospace Conference Proceedings (2017)
5. Baker, A.L., Keebler, J.R.: Factors affecting performance of human-automation teams. In: Advances in Human Factors in Robots and Unmanned Systems, vol. 499, pp. 331–340 (2016)
6. Billings, C.E.: Aviation Automation: The Search for a Human-Centered Approach. Lawrence Erlbaum Associates, Mahwah (1997)
7. Parasuraman, R., Sheridan, T.B., Wickens, C.D.: A model for types and levels of human interaction with automation. IEEE Trans. Syst. Man Cybern. Part A Syst. Hum. 30(3), 286–297 (2000)
8. Sheridan, T.B.: Adaptive automation, level of automation, allocation authority, supervisory control, and adaptive control: distinctions and modes of adaptation. IEEE Trans. Syst. Man Cybern. Part A Syst. Hum. 41(4), 662–667 (2011)
9. Endsley, M.R.: The application of human factors to the development of expert systems for advanced cockpits. Proc. Hum. Factors Soc. Ann. Meet. 31(12), 1388–1392 (1987)
10. Endsley, M.R.: Level of automation effects on performance, situation awareness and workload in a dynamic control task. Ergonomics 42, 462–492 (1999)
11. Brand, Y., Schulte, A.: Model-based prediction of workload for adaptive associate systems. In: Proceedings of the 2017 IEEE International Conference on Systems, Man, and Cybernetics, SMC 2017, pp. 1722–1727 (2017)
12. MIL-STD-2525C: MIL-STD-2525C: Common Warfighting Symbology. Changes, no. November, p. 1170 (2008)
13. Uhrmann, J., Schulte, A.: Concept, design and evaluation of cognitive task-based UAV guidance. J. Adv. Intell. Syst. 5(1) (2012)

Autonomous Ground Vehicle Error Prediction Modeling to Facilitate Human-Machine Cooperation

Praveen Damacharla$^{(\boxtimes)}$, Ruthwik R. Junuthula, Ahmad Y. Javaid,
and Vijay K. Devabhaktuni

EECS Department, College of Engineering, The University of Toledo,
MS 308, 2801 W Bancroft St., Toledo, OH 43606, USA
{Praveen.Damacharla,
RuthwikReddy.Junuthula}@rockets.utoledo.edu,
{Ahmad.Javaid,Vijay.Devabhaktuni}@Utoledo.edu

Abstract. Autonomous ground vehicles (AGVs) play a significant role in performing the versatile task of replacing human-operated vehicles and improving vehicular traffic. This facilitates the advancement of an independent and interdependent decision-making process that increases the accessibility of transportation by reducing accidents and congestion. Presently, human-machine cooperation has focused on developing advanced algorithms for intelligent path planning and execution that is functional in providing reliable transportation. From industry simulations to field tests, AGVs exhibited various mishaps or errors that have a probability to cause fatalities and undermine the potential benefits. Therefore, it is very important to focus on reducing fatalities due to either human error or AGV system error. To solve this problem, the paper proposes an error prediction model to reduce AGV errors through appropriate human intervention. In this paper, we use the data from AGV exteroceptive sensors such as stereo-vision cameras, long and short range RADARS, and LiDAR to predict the AGVs error through Dempster–Shafer theory (DST) based on sensor data fusion technique. The results obtained in this work suggest that there is a lot of scope for improvement in the performance of AGV when conflicts are predicted in advance and alerting human for intervention. This would, in turn, improve human-machine cooperation.

Keywords: Autonomous Ground Vehicle · Error prediction
Human-Machine cooperation · Dempster–Shafer Theory (DST)

1 Introduction

Currently, the Autonomous Ground Vehicle (AGVs) platform is witnessing an exponential rise in research and industry investment at the rate of 112% from 2010 to 2017 [1]. This tremendous growth warrants the future potential of AGV, and most of the upcoming innovations will be concentrated on semi-autonomous and autonomous vehicles. Reasons for this change are to reduce accidents due to human error, improvement in transport efficiency, free flow of traffic to the credit of quick reaction

© Springer International Publishing AG, part of Springer Nature 2019
J. Chen (Ed.): AHFE 2018, AISC 784, pp. 36–45, 2019.
https://doi.org/10.1007/978-3-319-94346-6_4

times, optimal fuel consumption and savings, and safe accessing of areas that are too dangerous to operate. There are many technological advancements that made the researchers and industries believe in the possibility of this future such as systems in a vehicle, highly efficient computation models, and cloud computing. However, yet with all these benefits there has not been a single fully autonomous vehicle realized so far, which is of level five autonomy as per SAE International J3016 Automation levels. The other four automation levels include driver assistant, partial automation, conditional automation and high automation respectively [2]. Among these, the level two partial automation is the only automation that has been implemented successfully on a commercial scale until now in Tesla Model S [2].

Around the globe, many research groups and industries such as Google, Uber, and Volvo have been working on various levels of AGVs. The testing of these vehicles, infield as well as in test circuits, has been in practice since 2007 and have recorded a sizable number of mishaps and errors. Some of the prominent mishaps/errors by AGVs are, Tesla Model S on June 30, 2016, operating in self-driving mode rammed into an 18-wheel truck, on Jan 22, 2018, Tesla Model S rammed into a fire truck that is parked on the 405 freeway in Los Angeles County, on February 14, 2016 [3], an autonomously driven Google Lexus collided with the side of a bus in Mountain View, California while preparing to turn right, in December 2016, Uber's computer-controlled car was caught on video running a red light, four seconds after the light turned red [4], in May 2016 while trying to demonstrate Volvo's pedestrian detection and auto-braking feature, the car automatically accelerated towards a group of people knocking them off their feet, in October 2016 NuTonomy self-driving cars hit a lorry at the one-north business park in Queenstown, and on May 11, 2015 [3], a self-flying car crashed during its test flight with the investor onboard. All major known errors and mishaps are summarized in Table 1.

Table 1. Error/mishap type and reasoning table

Date	AGV	Error/Mishap	Reason
February 14, 2016	Google Lexus	Car collided with the side of a bus	Car had detected the approaching bus, but predicted that it would yield and so is the driver
May 2016	Volvo	Car automatically accelerated towards a group of people	Pedestrian detection functionality and auto-braking failure
June 30, 2016	Tesla Model S	Car under self-driving mode rammed into an 18-wheel truck	Camera failed to recognize the white truck against a bright sky. Neither Autopilot nor driver hit the brakes
October 2016	NuTonomy	Self-driving cars hit a lorry	Speculations on auto pilot failure
December 2016	Uber	AGV in self-driving mode was caught on video running a red light	Speculations on AGV's camera and sensor failure along with human error
Jan 22, 2018	Tesla Model S		Speculations on autopilot failure and human error

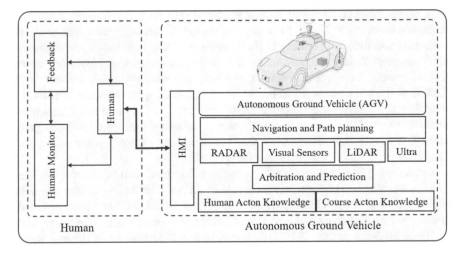

Fig. 1. Human Machine Collaboration Framework in AGV Operation

As shown in Table 1, these errors are a course by different systems of AGVs that are in testing on various levels of autonomy but almost all of them failed in preventing these errors before turning into an action. One of the primary reasons autonomous systems are used is to reduce accidents and human errors in the field. If AGVs had their own share of problems then it will undercut the primary purpose of the autonomous system implementation. Acknowledging the fact there will be no ideal system with errors reduced to 0% but there is still scope for error evasion through intervention. In that context, alerting the human promptly to a problem can prevent accidents due to AGV systems errors. A proper communication between the autonomous system and human with an action or intervention by the human can in this situation result in forming a human-machine teaming (HMT) or human-machine collaboration as shown in Fig. 1. There is no point of disagreement among the research community to endorse a fact that any active collaboration between human and machine can be translated into greater efficiencies compared to individual standalone task accomplishment or fully relying on either of them to finish a task [5]. Also, HMT can be defined as a system where human and machine have a two-way communication and both working on a common goal. For instance, an autonomous system operated by human in loop control will not be qualified as a fully autonomous system, according to published research studies a human-machine teaming will be a more trustworthy system than a fully autonomous system [5]. At current AGV operation, better human-machine cooperation can be realized if the human gets insights into errors/mishaps that an autonomous system might commit by a prediction model. Thus giving enough time for the human to intervene at the appropriate time to prevent any accidents.

In this paper, we are proposing a prediction method by using different types of data, in Sect. 2 we discussed systems in AGVs, related work, and background of prediction methods for different alert and prediction systems in AGVs. Section 3 focusses on various error models relevant to AGV error data for various sensors and presents an error prediction model through sensor data fusion using DST. Finally, Sect. 4 presents the results of the two test cases and is followed by conclusion in Sect. 5.

2   Background and Related Work

Different hardware systems in most common AGVs are stereo cameras sensors in front, side and back of a vehicle that monitors the nearby vehicles, lane markings, speed signs, and traffic lights etc. Long range and short range RADARs arranged at front and back of the car can monitor nearby vehicles and their relative motion and surrounding objects in streets. LiDAR is a sensor that emits LASER in 360°, which is affixed on top of the car and can monitor surrounding objects including vehicles. It works by generating point cloud data which is later used for 3D digital reconstruction of the environment. Low range ultrasonic sensors arranged all around the car can recognize low height objects. Wheel encoders, global position systems, conventional sensors such as have altimeters, gyroscopes, and tachymeters combined together to form an internal navigation system. Most of the research on AGV up to now have used cameras, RADARs, and LiDAR data with data fusion algorithms to identify dynamic environment surrounding the AGV. This information is then used to augment the AGV action or alert driver in the car [6–8].

In the real world, AGVs often practice warning and pre-alert commencements to alert human prior to wheel control handover as to avoid any possible mishaps and thereby provide safe navigation. For example, multitasking (phone conversing, chatting and video watching etc.) seriously affects human attention and therefore needs a proper alert system or warning mechanism to draw back human attention. Apparently, the alert configuration is categorized into brief alerts, pre-alert followed by a warning and repeated burst of pre-alerts [9]. All these alert mechanisms differ by response time allocated to human to take the wheel control. For an instance, a brief alert could be a typical voice warning followed by immediate transfer of control such as in Tesla Model S. However, the efficacy of brief alert system highly depends on human situational awareness. On the other hand, a pre-alert followed by an actual warning notice helps human respond quickly and facilitate quick restoration of situational awareness. Additionally, such an alert mechanism also improves human reaction rate [10]. Despite the aforementioned benefits, the efficacy of pre-alert system associated with a warning notice is still debatable entity among AGV research community and largely depends on degree of distraction. On this grounds, a repeated burst of audio pre-alert with varying time intervals could be more effective in restoring human attention regardless of degree of distraction.

However, no AGV has so far implemented an obvious error prediction model mechanism to commence intervention through either human or autonomous system [11].

2.1   Dempster Shafer Theory (DST)

The DS theory is a statistical framework related to Bayesian probability and deals with uncertainty, imprecise probability theories. While Bayesian statistics rely primarily upon combinations of internal probability factors within a system (such as propositions or random variables), DST is contingent upon external evidence factors, each one consisting of a range of data in which the desired output value can be found and a probability of confidence that the data range is in fact reliable. For example, in an AGV application, the DS theory can produce a variety of results relevant to the data supplied by sensor suit of AGV and plays a key role in developing an error prediction model which can be utilized in the alert system. The most important statistical output will be conflict value, a prediction of upcoming data is essential in determining alerts (this word choice can probably be improved or expanded). Other useful values include sensitivity, plausibility, and belief, as they can be used for reports that predict the human subject's overall health [12].

3   Prediction Strategy

3.1   Four Wheel Vehicle Error(s) Model

To successfully implement higher levels of autonomy, various engineering disciplines have to work together. Sensors collect the information about the vehicle's surrounding environment and is fused together to get information about the world around the car. This is generally referred to as perception. Navigation requires accurate location information and usually systems like Global Positioning System (GPS) and Inertial Measurement Unit (IMU) only, or sometimes the perception sensor information is taken into account as well. The decision making is done by some model by taking all this information and processing them to calculate the chance/risk of each action.

In practice, the models that calculate evidence, never cover all the situations, especially while operating in an unknown environment [13]. The error detection block considers this and builds on the DS theory. DS theory can measure the conflict between different sensors in forming a decision. Different reasons can cause conflicting belief values for different decisions.

3.2   Prediction Methodology Using DST

In this subsection, DST is introduced, with the main focus on the rules of combination and belief assignment that are used for the given application. DST is fed with the information required to make a decision, such as stopping at a red light, and it calculates the risk associated with the available options, which in this case are to either stop or proceed. The theory discussed below can be found in more detail in [14, 15], with other alternatives for combination rule and conflict management as a comparison of different sensor fusion algorithms is not the purpose of this work. The theory built on a mutually exclusive set of N world states given as

$$\Theta = \{w_1, w_2 \ldots w_N\}, \tag{1}$$

where Θ is called a Frame of Discernment (FOD) and every possible subset of Θ (power set 2^{Θ}) is assigned a belief value. For e.g. $\Theta = \{w1, w2\}$ will have its power set as

$$2^{\Theta} = \{\emptyset, w_1, w_2, \{w_1, w_2\}\}, \tag{2}$$

where $\emptyset$ is the empty set. Here mass functions and thereby belief functions are assigned to each member of the power set 2^{Θ} and are also comparable with the Basic Probability Assignment (BPA) in Bayesian theory [1]. However, DST implicitly includes the mass functions

$$m : 2^{\Theta} \rightarrow [0, 1] \tag{3}$$

The summation of mass function $m(A)$ over all possible subsets A in 2^{Θ} must be equal to one. DST uses additional measures to state the knowledge as a mass of a subset and is not necessarily related to a classical probability measure [1]. Three measures Belief ($Bel(X)$), Plausibility ($Pl(X)$) and Conflict ($Con(X,m1,m2)$) are defined, where X is the subset to be evaluated

$$Bel(x) = \sum_{A \in 2^{\Theta} | A \subseteq X} m(A) \tag{4}$$

$$Pl(X) = 1 - \sum_{A \in 2^{\Theta} | A \cap X = \emptyset} m(A) \tag{5}$$

$$Con(X,m1,m2) = \log\left(\frac{1}{1-(m1\ m2)(\emptyset)}\right) \tag{6}$$

$Bel(X)$ and $Pl(X)$ define lower and upper bounds for the classical probability interpretation

$$P(X) : Bel(X) \leq P(X) \leq Pl(X) \tag{7}$$

The conjunction formula $(m1 \cap m2)(X)$ is an important part of DS combination rule. This rule is used for combining measurements from different data sources and is defined as follows:

$$(m1 \cap m2)(X) = \sum_{AB \in 2^{\Theta} | A \cap B = X} m1(A) \cdot m2(B) \tag{8}$$

where masses $m1$ and $m2$ represent two different knowledges, for example, the knowledge about the environment from a camera sensor and Lidar sensor. A combined mass m_{12} results from fusing the two mass sets $m1$ and $m2$:

$$m_{12}(X) = \begin{cases} \frac{(m1\ \cap m2)(X)}{1-(m1\ \cap m2)(\emptyset)} & X \neq \emptyset \\ 0 & X = \emptyset \end{cases} \tag{9}$$

4 Results and Discussion

We have considered two of many hypothetical situations where the autonomous driving units have failed to take a right decision due to various environmental interferences and calibration issues that caused errors. We discuss those 2 scenarios and will discuss how DST can be used to come to final decision. In this paper, we also included an option of rejecting a decision altogether and switching control to the human user.

When switching control it will slow the vehicle down to give the user more time to react, hence making it feasible to implement in the real world.

4.1   Scenario 1

Let's consider a scenario where a car needs to stop at a signal since the light is red, and there might be a lot of environmental interferences like tail lights, street lights, hoardings with similar color etc. Let's further assume that there are 2 camera sensors for detecting the signal. Now due to the placement of the sensors, the environmental interferences would vary and hence result in very different probabilities for red, green, and yellow, which will be our world states. Therefore our power set would be

$$2^{\Theta} = \{\emptyset, \text{Stop}, \text{Go}, \{\text{Stop}, \text{Go}\}\}$$

In our case the subset {Stop, Go} is equivalent it Human Intervention (HI), because it implies a conflicted decision, it is the case when the sensor detects {yellow or green} or {red or green} together or it detects a light but not sure if it's the signal. The mass, belief, and plausibility functions are given in the table below, which are calculated using Eqs. (4), (5), (6), (8) and (9).

4.2   Scenario 2

Let's now consider a scenario where a car passing an overhead bridge on the highway. It requires readings from LiDAR and camera sensor to identify the obstacle and stop if there is an obstacle. The object detection is done by combining Lidar and camera sensors readings. Environmental interference, in this case, can be ground, overhead bridges, etc., might affect the readings of sensors, there might also be some calibration error that usually creeps in overtime. The power set is given by

$$2^{\Theta} = \{\emptyset, \text{Stop}, \text{Go}, \{\text{Stop}, \text{Go}\}\}$$

In our case, the subset {Stop, Go} is equivalent to Human Intervention (HI), because it implies a conflicted decision. The default cannot be set to stop in case of high conflict because it might result in unnecessary pit stops. The mass, belief, and plausibility functions are given in the table below, which are calculated using Eqs. (4), (5), (6), (8) and (9).

5 Discussions

The results are tabulated in Tables 2 and 3 and represent the mass functions obtained from different, sensor and combined mass functions obtained by using DS combination rule and also measure of conflict obtained from the conflict formulation of DST. These measures quantify the amount of uncertainty and thus helps us determine when it is better to have human intervention. We wouldn't want to bring the human into the picture too often as it would defeat the purpose of autonomous driving. DST often does a good job at combining evidence from two different sources, but it is shown to give counter-intuitive results in presence of high conflicts. We would calculate the conflict and decide whether to go with DST or bring inhuman. The hypothetical situations where a DST would typically fail to come with intuitive decision is shown, and situations like these have actually happened. Hence to have a framework to detect errors and switch to manual control will go a long way in improving safety of these applications. In both cases, the readings from the sensor had a conflict of 0.56 and **1.66,** which are very high, the first case would result in skipping the signal, and the second case would result in stopping unnecessarily in the middle of the highway. Therefore when the conflict is very high, the car would slow down and give the user the control, instead of implementing the decision.

Table 2. Individual and combined, mass, belief, and plausibility values for Scenario 1

	m1(X)	B(X)	Pl(X)
Stop	0.6	0.4	0.8
Go	0.05	0.2	0.6
Stop or Go	0.35	1	1
	m2(X)	B(X)	Pl(X)
Stop	0.05	0.47	0.8
Go	0.9	0.2	0.53
Stop or Go	0.05	1	1
	m12(X)	B(X)	Pl(X)
Stop	0.374	0.374	0.496
Go	0.504	0.504	0.626
Stop or Go	0.122	1	1

Table 3. Individual and combined, mass, belief, and plausibility values for Scenario 2

	m1(X)	B(X)	Pl(X)
Stop	0.9	0.47	0.8
Go	0.01	0.2	0.53
Stop or Go	0.09	1	1
	m2(X)	B(X)	Pl(X)
Stop	0.1	0.35	0.7

(continued)

Table 3. (*continued*)

	m2(X)	B(X)	Pl(X)
Go	0.7	0.3	0.65
Stop or Go	0.2	1	1
	m12(X)	B(X)	Pl(X)
Stop	0.499	0.47	0.8
Go	0.478	0.2	0.53
Stop or Go	0.023	1	1

6   Conclusion and Future Work

In this paper, we discuss how current autonomous vehicles have failed in many scenarios due to human errors and sensor errors. The lapses in technology prevent us from attaining full-scale measurement. Presently, human-machine cooperation is the best way to deploy the autonomous driving systems. To this end, we propose an error prediction model to reduce AGV errors through appropriate human intervention. We use the data from AGV exteroceptive sensors such as stereo-vision cameras, long and short range RADARS, and LiDAR to predict the AGVs error through DST based on sensor data fusion technique. The results obtained from the hypothetical scenarios we considered indicates that there is a lot of scope for improvement. This study can be extended to various scenarios where there might be high conflict and deciding a threshold for conflict value can be a direction for future research. In future we wish to implement study in a dSPACE simulator and test the effect of human intervention in human factor studies.

Acknowledgments. The University of Toledo and Round 1 Award from the Ohio Federal Research Jobs Commission (OFMJC) through Ohio Federal Research Network (OFRN) fund this research project; authors also appreciate support of the Paul A. Homer Family CSTAR (Cybersecurity and Teaming Research) Lab and EECS (Electrical Engineering and Computer Science) Department at the University of Toledo.

References

1. Gao, P., Kaas, H-.W., Mohr, D., Wee, D.: Automotive revolution–perspective towards 2030 How the convergence of disruptive technology-driven trends could transform the auto industry. Advanced Industries, McKinsey & Company (2016)
2. Favarò, F., Eurich, S., Nader, N.: Autonomous vehicles' disengagements: trends, triggers, and regulatory limitations. Accid. Anal. Prev. **110**, 136–148 (2018)
3. Noy, I.Y., Shinar, D., Horrey, W.J.: Automated driving: safety blind spots. Saf. Sci. **102**, 68–78 (2018)
4. Brenner, W., Herrmann, A.: An overview of technology, benefits and impact of automated and autonomous driving on the automotive industry. In: Digital Marketplaces Unleashed, pp. 427–442. Springer, Heidelberg (2018)

5. Chen, J.Y.C., Lakhmani, S.G., Stowers, K., Selkowitz, A.R., Wright, J.L., Barnes, M.: Situation awareness-based agent transparency and human-autonomy teaming effectiveness. Theor. Issues Ergon. Sci. **19**(3), 259–282 (2018)
6. Litman, T.: Autonomous vehicle implementation predictions. Victoria Transport Policy Institute (2017)
7. Shi, W., Alawieh, M.B., Li, X., Yu, H.: Algorithm and hardware implementation for visual perception system in autonomous vehicle: a survey. Integr. VLSI J. **59**, 148–156 (2017)
8. Dang, L., Sriramoju, N., Tewolde, G., Kwon, J., Zhang, X.: Designing a cost-effective autonomous vehicle control system kit (AVCS Kit). In: AFRICON, pp. 1453–1458. IEEE (2017)
9. Van der Heiden., R., Iqbal, S.T., Janssen, C.P.: Priming drivers before handover in semi-autonomous cars. In: CHI Conference on Human Factors in Computing Systems, pp. 392–404. ACM (2017)
10. Markoff, J.: Google cars drive themselves, in traffic. N. Y. Times **10**(A1), 9 (2010)
11. Johns, M., Sibi, S., Ju, W.: Effect of cognitive load in autonomous vehicles on driver performance during transfer of control. In: 6th International Conference on Automotive User Interfaces and Interactive Vehicular Applications, pp. 1–4. ACM (2014)
12. Elkin, C., Kumarasiri, R., Rawat, D.B., Devabhaktuni, V.: Localization in wireless sensor networks: a Dempster-Shafer evidence theoretical approach. Ad Hoc Netw. **54**, 30–41 (2017)
13. Högger, A.: Dempster Shafer Sensor Fusion for Autonomously Driving Vehicles: Association Free Tracking of Dynamic Objects (2016)
14. Liu, W.: Analyzing the degree of conflict among belief functions. Artif. Intell. **170**, 909–924 (2006)
15. Schubert, J.: Conflict management in Dempster-Shafer theory by sequential discounting using the degree of falsity. In: IPMU (2008)

Enhanced Human-Machine Interaction by Fuzzy Logic in Semi-autonomous Maritime Operations

Bjørn-Morten Batalden, Peter Wide$^{(\boxtimes)}$, Johan-Fredrik Røds, and Øyvind Haugseggen

UiT The Arctic University of Norway,
Klokkargårdsbakken 35, 9037 Tromsø, Norway
{Bjorn.Batalden, Peter.Wide}@uit.no,
{jro038, oha013}@post.uit.no

Abstract. Advanced autonomous maritime operations are today an emerging academic field, where the implementation of autonomous or semi-autonomous control, support and maintenance systems. The semi-autonomous operations often require a complex interaction between human knowledge and experience as well as suitable intelligent based programs. In this simulated approach of a ship's berthing operation, the captains' experience and knowledge is the basis for training the fuzzy logic system. The human-machine interaction can further be enhanced by a second fuzzy logic system to feedback the out-put fuzzy logic signal and adjust the berthing maneuver to find near-optimal solutions. The paper will present an Artificial Intelligent based semi-autonomous solution in maritime operations and discuss the related human factors as well as the sensors needed to define the decision support system for ship berthing operation and demonstrating by a proposed fuzzy logic-based solution.

Keywords: Artificial intelligence · Human-systems integration
Enhanced fuzzy logic · Maritime operations · Autonomous ships
Decision support systems

1 Introduction

In today's intelligent transportation systems, the evolving technologies in the emerging maritime sector [1], as well as in the present developments in autonomous cars [2] and Arial vehicles [3] is attracting great interest from both academia and industry. New intelligent solutions play an important role to achieve safer operations with better performance. Artificial Intelligent systems, like machine learning, fuzzy systems, and deep learning, can play an important role in many application scenarios. Towards the design of autonomous maritime operations, we can foresee stepwise progress towards semi-autonomous systems before designing fully autonomous solutions. The semi-autonomous maritime operations are based upon human-system approach, and an attractive trend is the decision support concept that is beginning to evolve [4, 5]. These support systems will provide enhanced human-machine interaction and merge the human experience and physical parameters into an integrated system. The decision

© Springer International Publishing AG, part of Springer Nature 2019
J. Chen (Ed.): AHFE 2018, AISC 784, pp. 46–54, 2019.
https://doi.org/10.1007/978-3-319-94346-6_5

support systems intend to provide safer and more accurate feedback by combining the human experience with sensor data and other information available.

The present maritime activities are mainly using the human capability in interacting with support systems, e.g., radar, AIS, etc. To handle complex operations, there is a need to merge with system intelligence to get decision making on the level of operating systems. The human is expected to handle both in planned operational activities as monitoring long and short-term behaviors, and operations as well as to make complex decisions of the overall system conditions, but also for unforeseen activities as, e.g., unnormality, malfunctions, and alarms. However, the human-machine interacting system of today has obviously reported weakness. When a failure occurs, then the human failure is obvious. Human failure is supposed to lie behind as much as up to 80% of the maritime accidents [6]. Therefore, stronger technology support is expected to have a major impact on performance in general and specific on safety issues.

When compared with the process of the state-of-the-art in the automotive industry, there are some papers verifying similarity of autonomous maritime operations and ongoing concepts of intelligent driving.

A widespread concept of levels of autonomy [7] is used in the automotive industry as well as within aviation. The concept is defined in the figure below from Society of Automotive Engineers and National Highway Traffic Safety Administration.

The level of autonomy defined in the automotive sector can be directly implemented in autonomous maritime operations. As seen from the figure above a distinct change exist between level 2 and 3, where a distinction occurs from partial to conditional automation. The partial automation has still the human as having control of the final decision making while entering the thin line to level 3, conditional automation. At this level, the systems in the unit will take over the decisions and consequently, is responsible for all decisions thereof, as also can be seen in [1]. The proposed structure of a decision support system in an assistive berthing approach, as demonstrated in this paper will then consequently, according to Fig. 1, belongs to level 2.

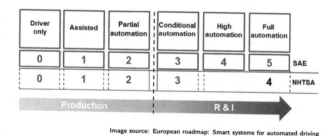

Image source: European roadmap: Smart systems for automated driving

Fig. 1. Level of autonomy [7].

In this paper, we will demonstrate the capability of the enhanced human-machine interaction in a berthing operation of a ship and the supportive action feedback to the crew on-board, to recommend the right decisions when moving into the harbor and entering the quay. The tests are performed by simulations in MathWorks and the FL-toolbox [8].

2   Human Factors in Maritime Operations

In maritime operations, the ability to make sound and safe decisions in a complex and dynamic environment requires a certain level of understanding of what is happening and what will happen. A framework suitable for analyzing this operational understanding is situation awareness (SA) [9]. The result of a good SA is the ability to predict future consequences of the current state and actions, taking into consideration several aspects of an operation. Endsley and Jones [10, p.13], define SA as "the perception of the elements in the environment within a volume of time and space, the comprehension of their meaning, and the projection of their status in the near future." In their study of maritime casualties and incidents, Batalden and Sydnes [12] find that 28 percent of the causal factors related to unsafe acts and that "decision and judgement errors" were ranked number 4 out of a total of 29 categories at third tier of the human factor analysis and classification system (HFACS) [11]. Better decision support systems together with other safety measures might reduce the decision and judgment errors and enhance other dimensions of HFACS.

Endsley and Jones [10] present SA comprising of three levels, (1) perception of the elements in the environment, (2) comprehension of the current situation, and (3) projection of future status. From an SA perspective, a navigator onboard a ship will use his eyes and ears to perceive important elements in the environment and from instruments. Looking at the sea and listening to the wind will inform the navigator in addition to reading instruments. In the aviation industry, Jones and Endsley [13] report that 76% of all SA errors among pilots related to not perceiving the needed information (an error at level 1). It is not proven that this number applies to the maritime domain, but it is not unlikely that certain maritime operations, such as berthing a vessel, are subject to similar problems. Facilitating necessary information to maintain a good SA during berthing seems central to improve the reliability of human performance.

2.1   The Berthing Operation

This simulated approach will demonstrate the abilities and improved performance when operating a ship into the harbor by using an assisted system in interaction with the crew, as described in Fig. 2. The fuzzy logic supportive system will increase the performance of the operation and improve the traditional human experience process by integrating the human experience of berthing procedure with a human-based intelligent fuzzy logic approach of an assistance system, The system uses a human experience based set of membership functions and rules that have been implemented in the support system. The enhanced human-machine interaction will make use of crew experience as input, as seen in Fig. 2. The enhanced process will be data-driven and compensate for new human decisions and experience. This means that the output from the original berthing approach system will be used as input for the second part of the berthing approach assistance system. The aim for the first part of the berthing approach system is to identify the ideal course for berthing approach under different conditions. The aim of the second part of the berthing approach system is to adjust the actual vessel course to achieve the best adjustments in relation to the ideal course from the first part of the system.

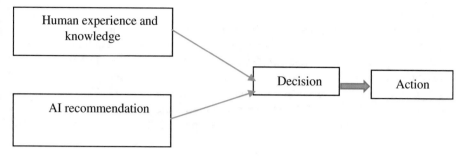

Fig. 2. The human – machine interaction model.

To summarize, the following is used as input for the second part of the system:

- Ideal Course from the first part of the berthing assistance system
- Actual vessel course

This leads to the following output:

- Course adjustment needed for achieving the best-enhanced adjustments in relation to the ideal course.

By implementing this second part of the berthing approach system, the overall approach can be performed autonomous and implemented in an enhanced fuzzy logic design, as can be seen in Fig. 3.

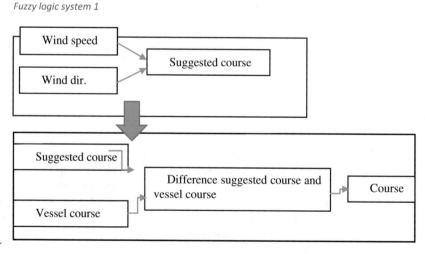

Fig. 3. The enhanced fuzzy logic model.

2.1.1   Simplifications

The simplifications and limitations from the first part of the system are still in force. In addition to these simplifications, it is necessary to add one simplification related to the second part of the system. This simplification is related to the fact that the system is depending on a pier with infinite length, which is not the case in reality where the vessel is supposed to moor at a designated spot on the pier, as can be seen in Fig. 4. For further research, it is preferable to add another step to the berthing approach system, which could adjust the vessel course in several steps to approach the pier at a designated spot.

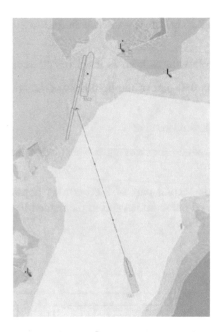

Fig. 4. The simulated berthing operation.

3   Decision Support System

The aim of the fuzzy logic berthing approach is to use this system as a decision support system for the navigational officers when executing complex berthing operations. The use of a decision support system is expected to increase the performance of the crew and make complex maritime operations safer. The crew on-board will gain the possibility to take in account a calculated feedback recommendation to increase the performance and as a result most likely reduce the risk for an accident causing human and material damage, as shown in Fig. 2.

3.1   Human-Based Knowledge and Experience

For berthing operations in general, much of the navigational officers' competence is acquired through experience from previous operations, often executed under different conditions. The accumulated knowledge and experience acquired competence can be used to present and support less experienced crews without the need for other human interaction and will provide them with alternative solutions to the complex situation assessed. If the system is used right and is making use of input of adequate quality, it can even increase the safety, as the probability of human error and breakdown in bridge communication is reduced. Such a decision support system built upon human-based knowledge and experience is also very interesting in an autonomous way of thinking. On a fully autonomous vessel, there will not be any navigational officers present to assist in complex maritime operations with human-based knowledge and experience. It is therefore of interest to make this information available for the vessel through the use of advanced technology, such as this decision support system.

3.2 Simulation by Fuzzy Logic Toolbox in MathWorks

The simulation was made in MathWorks Fuzzy logic toolbox with fuzzy rules that have been allocated in coherence with the human way of performing the berthing operation. The crew's earlier experience and nautical knowledge are transferred to the fuzzy logic system 1 for an ideal route and to the fuzzy logic system 2 for finding the optimal route in the berthing operation, as shown in Figs. 5 and 6 below.

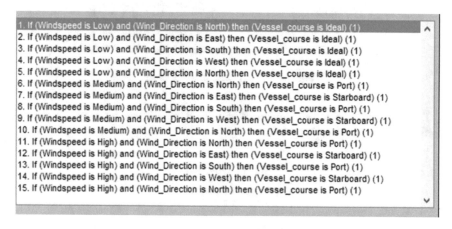

Fig. 5. Rules fuzzy system 1

The simulation uses the three input parameters wind direction, wind speed and vessel course as the human perception as input to the fuzzy system 1. By interacting the human experience of nautical knowledge and earlier experience then the crew can relate the input wind parameters as low, medium and high and weather directions, i.e., east, west, north or south than an action is taken, as seen in Fig. 8.

At low wind speeds, the direction of the wind does not lead to a change in the output-course as indicated in Fig. 7. At higher wind-speeds, a significant course-change can be seen when the wind direction changes. A correlation can be seen between opposite wind-directions if the vessel course is constant. For wind-speed

1. If (Ideal_Course is Ideal_Course_1) and (Actual_Course is Actual_Course_1) then (New_Course is New_Course_1) (1)
2. If (Ideal_Course is Ideal_Course_2) and (Actual_Course is Actual_Course_1) then (New_Course is New_Course_2) (1)
3. If (Ideal_Course is Ideal_Course_3) and (Actual_Course is Actual_Course_1) then (New_Course is New_Course_3) (1)
4. If (Ideal_Course is Ideal_Course_4) and (Actual_Course is Actual_Course_1) then (New_Course is New_Course_4) (1)
5. If (Ideal_Course is Ideal_Course_5) and (Actual_Course is Actual_Course_1) then (New_Course is New_Course_5) (1)

Fig. 6. Rules fuzzy system 2

above 15 m/s, the system will recommend no course, meaning that berthing is not recommended under such conditions.

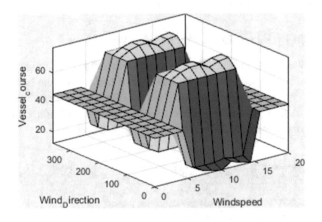

Fig. 7. The resulting mesh diagram.

Tables 1 and 2 states the value of the input parameters used as membership-functions in the fuzzy logic system.

Table 1. Wind speed.

Wind speed	Value [ms^{-1}]
Low	0–5
Medium	5–10
High	10–15
Extreme	15–20

3.3 Enhanced Fuzzy Logic System

The enhanced fuzzy logic system is based on the original approach of a fuzzy logic system which gave the user an ideal vessel course for approaching the pier based on the input parameters wind speed and direction. The improved system uses the output from the original system, meaning the ideal vessel course, as an input alongside with the actual vessel course. As an output from the improved system the users are presented

Table 2. Wind direction.

Wind direction	Value [^{0}]
North	315–45
East	45–135
South	135–225
West	225–315

with the course and the vessel needs to steer to approach the pier in an ideal way. By implementing the vessel's autopilot into this process, the berthing approach can also be made fully autonomous.

Figure 8 shows a vessel approach to berth at a given course and under given conditions. For the situation described above, the vessel course is 030°. The system recommends a course of 060° based on the initial conditions. This leads to a course change 30° towards starboard.

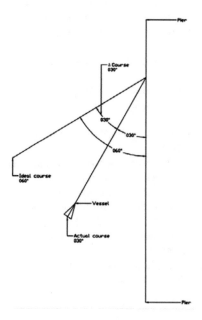

Fig. 8. The proposed enhanced fuzzy logic application.

4   Conclusion

In the long term, fully autonomous solutions will be possible to implement in maritime operations, and the development is expected to be performed in close connection with the maritime business, organizations and governmental authorities. Societal regulations and frameworks from maritime business, organizations and governmental rules will lay the foundation for smart developments of fully autonomous solutions that, for example, makes the crew able to control the ship from on-land command centers.

Further work will consist of improving the system by adding more steps, with the aim being able to dock at a pier with limited length. It is also preferable to implement vessel speed as a parameter to improve the berthing approach. It is also necessary to adjust the system for different types of vessels with different maneuver- characteristics and equipment.

References

1. Batalden, B.M., Leikanger, P., Wide, P.: Towards autonomous maritime operations. In: IEEE International Conference on computational intelligence & virtual environments for measurement systems and applications CIVEMSA (2017)
2. Guvenc, L., Guvenc, B.A., Emirler, M.T.: Connected and autonomous vehicles: Internet of Things and Data Analytics Handbook, pp. 581–595 (2017)
3. Valavanis, K.P., Vachtsevanos, G.J.: Future of unmanned aviation. In: Handbook of Unmanned Aerial Vehicles, pp. 2993–3009. Springer, Netherlands (2015)
4. Lughofer, E.: Evolving fuzzy systems-methodologies, advanced concepts and applications, vol. 53. Springer, Berlin (2011)
5. D'Urso, P., Massari, R.: Fuzzy clustering of human activity patterns. Fuzzy Sets Syst. **215**, 29–54 (2013)
6. Rothblum, A.M.: Human error and marine safety, U.S. coast guard research & development center. In: 2nd International Workshop on Human Factors in Offshore Operations, Houston (2002)
7. SAE, Artificial intelligence becomes a reality. 17 May 2017. http://articles.sae.org/15337/
8. MathWorks Fuzzy Logic Toolbox
9. https://se.mathworks.com/help/fuzzy/index.html?s_tid=srchtitle. 27 February 2018
10. Endsley, M., Jones, W.M.: Situation awareness information dominance & information warfare. Logicon Technical Services Inc., Dayton (1997)
11. Endsley, M., Jones, D.G.: Designing for situation awareness: an approach to user-centered design, 2nd edn. CRC Press, Boca Raton (2012)
12. Batalden, B.M., Sydnes, A.K.: Maritime safety and the ISM code: a study of investigated casualties and incidents. WMU J. Marit. Aff. **13**(1), 3–25 (2014)
13. Jones, D.G., Endsley, M.R.: Measurement of shared SA in teams: Initial investigation (No. SATech-02-05). SA Technologies, Marietta (2002)

Investigation of Unmanned Aerial Vehicle Routes to Fulfill Power Line Management Tasks

Gunta Strupka[✉] and Ivars Rankis

Institute of Industrial Electronics and Electrical Engineering, Riga Technical University, Azenes Street 12/1, 1048 Riga, Latvia
Gunta.Strupka@edu.rtu.lv, Rankis@rtu.lv

Abstract. Paper presents unmanned aerial vehicle (UAV) usage options and importance to fulfil power engineering tasks, improvements and their explanation. This paper also is part from research in this field restarted this year based also on previous research connected with power consumption problems [1] and propose initially data according topology without full coverage of data necessary to present final data.

Keywords: Human factors · Unmanned aerial vehicle · Power engineering Overhead line · Power consumption · Lidar · PLS-CADD · Expert systems Control systems

1 Introduction

In the recent years unmanned aerial vehicles have been widely researched and their possibilities of usage are rapidly increasing. In the previous Author's publications several usage possibilities of UAV were researched and analyzed [1].

A specified insight in problematic issues of UAV and increase of its possibilities became topical when the drone turned from an expensive toy into a companion for sportsmen and operator of any kind. Its lift capacity and speed has increased significantly [2].

In field of civil engineering and power engineering constructional stability is one of most significant checkpoints. When buildings become skyscrapers and power generation and transition systems must provide rapidly growing cities with energy and build new, bigger and much energized power lines, investigation of their foundation stability starts to be almost as significant as the power quality problem [3].

It is very important to investigate widely used public constructions like bridges to avoid any disasters. UAV also could be used for investigation of historical constructions and wide area objects [4]. But scans of substation during its construction is just a checkpoint in this article.

Due to the fact high voltage overhead lines (OHL) belongs to national objects they must be followed carefully but researches in this field are limited.

2   Power Line Recordings and Insight of Recognition Process

In power engineering field are lots of parameters that must be managed not only during construction, but also during all further operation. Not only type of power line is important and its main elements like towers, wires and its technical accessories like insulators, armature, but also its geopositional gauges – sags of wires, gauge till ground or any object in range. Not only all mentioned before are important to ensure secure and stable power system, but condition of tower or pole [5], which becomes the main theme of this article.

There are already some good methods that are used to fulfil this task, but in the same time lack of option to do it automatically. For example, very good and precise results are reached from LIDAR scans [6] that are good to implement in other featured software like PLS-CADD [7] and then be analyzed. All voluminous data are transferred from scanner on field to hardware at office that makes it more time consuming and raise risk to lose data if something happen with scanner accessories.

In the same time at office data are analyzed mostly via visual detection by experts [8]. And it is based on very wide experience and knowledge, but also time consuming and not so effective for long distance power lines.

To prevent risky and time consumption factors, tower detection and gauge calculations could be done automatically, and it would be performed all over the power line and all calculated data could be saved also about normal constructions for usage in future in case of surveys during long period of time. It should be recalled that expert knowledge is priceless for precise reaction during uncommon situation.

Another importance of such scans during building works of new power line and its connection to existing one. As such designs must be calculated very carefully (span lengths, sag tensions, structure forces) and it is done on whole power line from one substation till next, so designers must be sure about its stability and technically possible forces on each constructional element [5].

2.1   Overhead Power Line Visual Detection

Automate visual detection of tower could be performed via artificial intelligence like deep learning pattern recognition. Main point is to understand how tower structure is build.

Normally tower consist of four fundaments located on square corners, tower crated structure, cross-bar and armature with insulators, Fig. 1(a). During some specific projects towers were created in very different ways like deer, bug, human figures without only common element as armature.

After managing tower structures over long EPL, it is time to recognize full construction of towers, Fig. 1(b).

From mentioned above figures it is seems quite easy to catch main similarities common for all power line constructions.

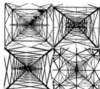

Fig. 1. Tower (a) construction projections; (b) images after filters.

2.2 Important OHL Parameters

Within OHL structure detection automation, list of important preplanned parameters should be noticed. Mostly they are connected with safety tasks. Theoretical knowledge and improvements of UAV automation tasks will be useless if physical parameters of explored object will not be taken into account. Such parameters can't be changed if software will need it.

Most significant is to understand height of tower, its dimensions, next are wires and line armature. And then expert knowledge about OHL constructions and their performance must be involved. Beside lots of safety parameters and factors must be included started from overall standards about OHL construction parameters [9] and construction standard of power line locations and distances from other constructions [10] till safety requirements for working under voltage [11].

Mentioned parameters from Table 1 will be used as minimal distance in next calculations.

Table 1. Allowed distance to power leading parts [11].

OHL voltage	Distance from human, instruments and equipment (m)
110 kV	1,0
330 kV	2,5
Lighting protection wires	1,0

3 UAV Power Consumption According to Its Route Over OHL

After understanding of power line constructions and their location its time to start planning and analyzed UAV improvements.

Important research must be done with AI assistance during scanning, that could allow quick checks and overlooking in long distance and perform much deeper investigations in dubious places. After finishing recognition training set, many critical stone points must be implemented according to power engineering rules to make UAV into autonomous and energy efficient tool.

3.1 UAV Mathematical Model

Mathematical model of UAV should be created. This model consists of its target function [1].

$$F_{UAV} = \begin{cases} E_{UAV} = \int_i^n UIdt \rightarrow min \\ t_{UAV} = f(E_{UAV}, P) \rightarrow max \\ E_{battery} > E_{base} + E_{add} \\ E_{batRez} \geq 15..20\% of E_{UAV} \\ F_v = F_T - mg \\ F_a = c_a \frac{\rho v^2}{2} S \end{cases} \quad (1)$$

Equation (1) presents some of important functions and tasks:

- E_{UAV} – consumed electrical energy by UAV(W);
- t_{UAV} – working duration (s);
- $E_{battery}$ – battery energy capacity (W);
- E_{base} – energy consumption used by main devices (W);
- E_{add} – energy consumption of added devices (W);
- E_{batRez} minimal battery reserve to make battery work longer (W);
- F_v – vertical force (W);
- F_T – thrust (g/W);
- F_a – aerodynamic drag (g/W);
- c_a- aerodynamic drag factor;
- ρ - air density (kg/m^3);
- v – speed of air flow (m/s);
- S- the surface/area of vehicle acted up by air flow (m^2).

After main mathematical model is created, initial parameters of UAV for further calculations should be considered, Table 2. After initial input data and technical parameters are set [12], it could be optimized in next table.

Table 2. UAV mechanical and electrical data.

Drive Weight	1848	g	Full Weight	2260	g
Thrust-Weight	2.4:1				
Hover			Max load		
Current	25.89	A	Current	103.24	A
P(in)	383.2	W	P(in)	1528	W
P(out)	303.1	W	P(out)	1208.1	W
Efficiency*	79.1	%	Efficiency*	79.1	%
Max Speed	16.67	m/s			
Estimate rate of climb	11.1	m/s			

*Efficiency percentage is calculated by online calculator and presents only rate for idealized hover state and max load (climbing, vertical take-off).

UAV take-off trajectory [13] is vertical till its cruise height. No changes in vertical location will be discussed, path planning only in 2D plan. Test route length is 1 km. For safety reason it is important to follow protection zones of OHL [14].

3.2 Straight Route

Overhead line route in straight line is the most welcome but not always possible. But when it takes place its parameters are relatively easy to calculate in any section.

There is no need to calculate many possible options and tracks for UAV over such line, as it will not have any changes depending on UAV location on the left or right side of OHL (Fig. 2).

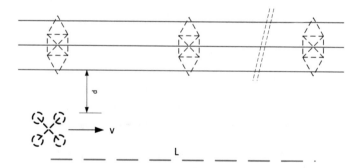

Fig. 2. Straight route schematic.

Near point 1 as seen in Fig. 3, UAV takes off at max load with speed of 11.1 m/s. Route between 1 and 6 in total length 1 km is done within 60 s. Landing take down is done with on quarter reduced speed and power. Total power consumption is 26,4 kW.

Of cause, such speed is impossible to reach during scanning process as for certain camera or scanner technology it is important to fill stability requirements.

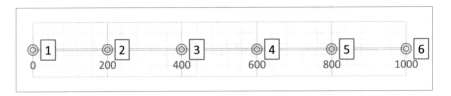

Fig. 3. Straight route locations and tower numbers.

3.3   Route with Turning to One Direction

Route with turnings consists not only from anchor and suspension towers but also of angle suspension towers or sometimes with anchor tower instead (Fig. 4).

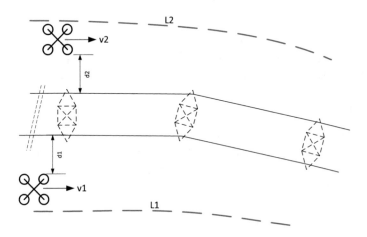

Fig. 4. Schematic of route with turning to one direction.

Mathematically it is understandable that path in narrow angle side (L1) is shorter then wide angle side (L2).

As in case of straight route takeoff and landing are under similar conditions but point 1 to 6 is in hover mode. Only difference is near point 3, Fig. 5, where UAV should change its direction for some degrees. Within this example in 40-degree angle is used.

As result right side route (narrow angle) is shorter then left side route that express in difference of more than 3 kW to one or another side.

3.4   Route with Multiple Turnings

Overhead lines outside short viewed sections as in previous examples are always complex with all possible combinations included (Fig. 6).

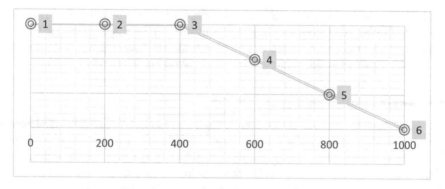

Fig. 5. Route with turning to one direction locations and tower numbers.

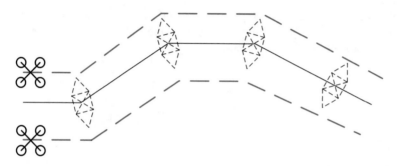

Fig. 6. Schematic of route with multiple turnings.

In this route version will be more than two possible trajectories of flight (Fig. 7).

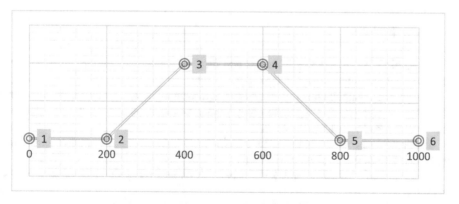

Fig. 7. Route with multiple turnings locations and tower numbers.

Initially trajectories will be calculated in previous example with one turning on both sides. And in long distances such trajectories in sum will be almost the same. Left side flight also takes more power because of longer path, but difference was less then 2 kW on the same rout as single angled route. Also, angles are the same.

3.5   Route Path Planning Improvements

It is not enough to consider of using left or right side of OHL to inspect it. To improve power consumption and make it more adaptive additional movements will be done.

Additional researches should be done in a field of electromagnetic influence on UAV during flight over OHL as its field could be used to prolong flight time or decrease it dependently from distance between UAV and voltage sources.

As mentioned before, distance plays big role in power consumption and safety. It should be optimized.

Situation as in Fig. 8 is also possible, but all previous rules must be taken into account. To fly bellow power line its sags must be known, but when fly above not always all data are possible to collect.

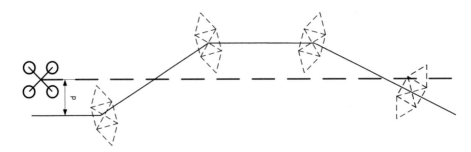

Fig. 8. Schematic of improved UAV route path over multiple angle OHL.

4   Conclusions

The possible UAV routes are discussed and their parameters are defined, which allow to consider operational parameters of UAV.

Within these calculations lots of additional information must be taken into account. One of most significant is regulation of overhead power lines about allowed distances and dimensions to ensure safety.

Angle of inspectable route makes impact on UAV power consumption and route flight must be preplanned more detailed.

Path planning improvements keep pace with pattern recognition process to avoid disasters.

During further research UAV will be able to collect set of initial rules and optimize its path to fill power engineering tasks more detailed and longer as it is possible as it is now. Additional researches should be done in a field of electromagnetic influence on UAV during flight over OHL.

This paper presents necessity to upgrade existing preplanned route plans into highly improved autonomous software that understands mission and its conditions, compile its own route and dealing with its own power consumption problems.
Work of research will continued.

References

1. Strupka, G., Levchenkov, A., Gorobetz, M.: Fuzzy-logic algorithm of UAV hardware configuration assessment for flight time and lift capacity improvements. In: 58th International Scientific Conference on Power and Electrical Engineering of Riga Technical University (RTUCON) (2017)
2. Aerones. https://www.aerones.com/eng/
3. Strupka, G.: Algorithm for unmanned aerial vehicle to supervise applications for civil and power engineering tasks. In: 17th International Symposium Topical Problems in the Field of Electrical and Power (2017)
4. Chen, S.-E.: Laser scanning technology for bridge monitoring. In: Apolinar Munoz Rodriguez, J. (ed.) Laser Scanner Technology. InTech (2012)
5. Lvov, A., Priedite-Razgale, I., Rozenkrons, J., Kreslins, V.: Assessment of different power line types' life-time costs in distribution network from reliability point of view. In: 2012 Electric Power Quality and Supply Reliability, Tartu (2012)
6. Thakur, R.: Scanning LIDAR in advanced driver assistance systems and beyond: building a road map for next-generation LIDAR technology. IEEE Consum. Electron. Mag. 5(3), 48–54 (2016)
7. PLS-CADD (Power Line Systems - Computer Aided Design and Drafting). http://www.powline.com/products/pls_cadd.html
8. Aerial Laser Scanning. http://www.optensolutions.com/cntnt/eng/survey.html
9. Gaisvadu elektropārvades līnijas ar spriegumu 110–330 kV (in Latvian). https://www.latvenergo.lv/files/text/energostandarti/LEK_135.pdf
10. Noteikumi par Latvijas būvnormatīvu LBN 008-14 "Inženiertīklu izvietojums" (in Latvian). https://likumi.lv/doc.php?id=269200
11. Drošības prasības, veicot darbus uz 110–330 kV elektrolīniju daļām, kuras ir zem sprieguma (in Latvian). https://www.latvenergo.lv/files/text/energostandarti/LEK_096.pdf
12. xcopterCalc - Multicopter Calculator. https://www.ecalc.ch/xcoptercalc.php
13. Strupka, G., Levchenkov, A., Gorobetz, M.: Influence of take-off trajectory on quadcopter energy consumption. International Scientific Conference Transport Means (2017)
14. Protection zone law. http://www.likumi.lv/doc.php?id=42348

Information Displays and Crew Configurations for UTM Operations

Quang V. Dao[1]([⊠]), Lynne Martin[1], Joey Mercer[1], Cynthia Wolter[2], Ashley Gomez[2], and Jeffrey Homola[1]

[1] NASA Ames Research Center, Moffett Field, CA, USA
{quang.v.dao,lynne.martin,joey.mercer,
jeffrey.r.homola}@nasa.gov
[2] San Jose State University Research Foundation,
210 N 4th Street, San Jose, CA, USA
{cynthia.wolter,ashley.n.gomez}@nasa.gov

Abstract. In this paper we discuss how team configuration may influence how information is shared among team members for low-altitude Unmanned Aircraft Systems (UAS) operations. NASA collected and analyzed observation data gathered during a series of field tests for the UAS Traffic Management (UTM) project. The field tests were part of a larger effort aimed at advancing the UTM concept, conducted at six test-sites across the USA. Ground control station (GCS) concepts, flight-crew composition, and crew-size varied within and across test-sites. Flight crews took two strategic approaches to organizing their teams. The first of the two approaches was implemented by one third of the flight crews. These crews integrated the role of UTM operator into the duties of existing crew members, merging the current roles with this new one, keeping the UTM operator collocated with the flight crew. The remaining two thirds implemented a distributed team configuration, where a single UTM operator distributed support across multiple crews. Results from our data collection efforts revealed that UTM operator location influenced whether flight crews used verbal communication versus displays to acquire UTM information.

Keywords: UTM · UAS · Teams · Situation awareness

1 Introduction

The focus of this paper is on the impact of unmanned aircraft system (UAS) crew configuration on how information pertinent to a crew's objectives is used and accessed. These crews operate unmanned aircraft (UA) for diverse applications. They may fly the aircraft manually or provide high level direction if the aircraft is autonomous. We expect that most UA traffic will be autonomous because the quantity of aircraft – particularly for major commercial operations – will exceed the human operator's bandwidth for actively managing the flights. By 2035, these commercial operations will comprise more than half of the projected 250,000 concurrent UAS operations [1]. The overall UA fleet size is forecast to be 35 times the number of manned aircraft currently in operation. To meet this expected demand, NASA is currently working with

© Springer International Publishing AG, part of Springer Nature 2019
J. Chen (Ed.): AHFE 2018, AISC 784, pp. 64–74, 2019.
https://doi.org/10.1007/978-3-319-94346-6_7

academic and industry partners to develop a concept for a UAS Traffic Management (UTM) ecosystem [2]. The UTM system will manage traffic for unmanned aircraft that are less than 55 lbs, and only for operations occupying airspace up to 400 ft in altitude. At the center of this ecosystem is an information exchange system that allocates active management of aircraft to automation and provides services to UAS operators for coordinating shared access to airspace [3]. In a series of four flight test campaigns NASA will observe and analyze how UAS operators incorporate UTM tasks, displays, and tools with their crews [4]. These four campaigns will be supported by four distinguishable technical capability levels (TCL).

1.1 Flight Test Campaigns

The TCLs build on each other and will include tests that grow in complexity. TCL 1 and 2 have already been completed. TCL 1 was limited to operations within visual line of sight and no more than 2 concurrent operations. It served to test very basic UTM functionality, which included vetting proposed flights for lateral conflicts with active operations and pre-defined constraints such as airports. In TCL 2, beyond visual line of sight (BVLOS) operations were introduced, along with enhancements such as alerting for airspace intrusion and segmented flight planning, that vetted operations for both lateral and vertical conflicts so that concurrent airspace reservations can be stacked by altitude. TCL 3 and 4 will be conducted within two years from the writing of this paper. They will include operations over increasingly populated areas, between moderate and high UAS traffic densities, interactions between manned and unmanned operations, as well as large-scale contingency management. For a more detailed discussion of the flight campaigns and TCLs see Johnson et al. [4], and for a more detailed discussion of NASA's concept of operation see Kopardekar et al. [2]. The findings reported in this paper will focus on the most recent test - TCL 2.

In TCL 2, displays that provided information about the availability of mission airspace and information about where their aircraft were relative to operational boundaries, as well as other air traffic were introduced. How information was accessed depended on the configuration of the flight crew. This configuration was influenced by the role of the crew member responsible for UTM tasks and how s/he coordinated with the rest of the crew to disseminate UTM information for achieving mission objectives. In the next section we stage our analysis of the UTM crew with a brief discussion about teams and point to distinct features necessary for them to function.

1.2 Share Cognition and Situation Awareness

In this document, we define teams to be UAS ground control station crews composed of two or more individuals who have distinguishable roles, but who are interacting interdependently to achieve a common objective [5]. Endsley [6] points out 3 major features of teams from this definition. The first is that team members share common objectives. Second, team members have specific role, and, third, the members are interdependent. Commonly, functional teams will collectively form a construct called shared cognition [7]. Shared cognition develops when team members share four specific types of information through effective communication and coordination:

(1) knowledge about a task, (2) knowledge of processes pertaining to a task, (3) knowledge about team members, and (4) their attitudes or beliefs [7]. Acquisition of the aforementioned types of knowledge allow the formation of mental models that are shared between members of a team. These models are important because individuals can then anticipate appropriate action in advance of verbal approval from others – particularly during periods of high workload.

As long as the appropriate types of information are being propagated, effective teams need only share enough information to reach a common understanding. Endsley and Jones [6] use the term shared situation awareness (SA) to refer to information that is actually shared among team members. For UTM, this information includes an aircraft's position relative to its operational boundary and surface obstructions. The purpose of shared SA is to achieve team SA. The key difference here is that team SA is achieved when team members share an understanding that is an accurate reflection of the relevant events and system(s) in the environment. It is not always the case that shared SA leads to team SA. For example, a UA ground control station operator monitoring a display may concur with a visual observer who is also tracking an aircraft in the sky that there is enough fuel capacity to complete a mission, but both team members may neglect to identify changing wind conditions that would accelerate the burn rate on the fuel cell, effectively requiring the aircraft to return sooner than planned. In this paper, we focus on how shared SA was achieved as a function of the UAS team (i.e., flight crew) configuration, and now turn to a description of the flight test environment in which these teams operated before discussing results from our data collection efforts.

2 Method

A series of flight tests were conducted at six different sites located across the USA, during May and June of 2017. The campaign resulted in over 240 data collection flights. Flights not only varied in duration, but also in the environments and terrains over which they flew. The flight tests highlighted beyond visual line of sight (BVLOS) and altitude-stratified operations. Scenarios were developed independently by each test-site to demonstrate the UTM capabilities that they had proposed, and were required to include some combination of BVLOS, altitude stratification and multiple vehicles in the air. Some test-sites created one scenario with a series of variations to capture these capabilities, while other test-sites constructed a number of unique scenarios. Test sites reviewed and modified their scenarios in one or more shakedown days, that commonly included equipment testing (e.g., testing connectivity with UTM), while data-collection days focused on completing flights meant to satisfy the test scenarios. More detailed numbers regarding how these flight activities were distributed over the six test-sites are available in Martin et al. [8].

2.1 Crew Roles and Responsibilities

A total of 23 flight crews participated in the tests. Flight crews varied in number and affiliation: some had just two individuals, while others had approximately twelve in their crews. Flight crews from some test-sites were composed of individuals from one organization, while other test-sites' crews had members from different organizations. Primary flight crew positions included those listed in Table 1 and additional positions staffed by some, if not all, of the flight test sites are listed in Table 2.

2.2   Crew Configuration

Although each test-site created its own crew configurations, two types of team organization emerged with respect to UTM – integrated and distributed. In integrated teams, organization included having the USS Op role (Table 1) as a dedicated member of the flight crew, either completing USS client management tasks alone, or by having one crew member splitting the USS Op role with another role (e.g., at Test-Site 3 the flight crew consisted of two people: a GCSO/PIC/USS Op and a safety pilot/launch engineer). The advantages of having the USS Op role within a flight crew team was that this person was able to focus completely on the crew's mission and communications were reduced. The cost was the number of additional personnel, or, if the role was time-shared by one team member, that periods of high workload were compounded if all roles were busy at the same time, (e.g., at take-off and landing).

In distributed teams, a dedicated USS Op fulfilled this role for a number of crews (e.g., at Test-Site 1, one USS Op submitted and managed the flight volumes for four flight crews, where each flight crew consisted of a PIC, a GCSO, and a launch engineer). Three test-sites took this "hub-&-spoke" approach – Test-Sites 1, 4 and 5. The advantage of separating out the USS Op role was that this person became a specialist and overall required manpower was reduced. The cost was the increase in communications load, as the USS Op had to stay in contact with all the flight crews s/he was serving, and the workload related to managing multiple flights in the case that one flight crew/vehicle was having an off-nominal event.

2.3   Tools and Displays

Equipment available at each Ground Control Station (GCS) location varied widely across (and sometimes within) test-sites. At most GCSs, several displays were available to the flight crews to give them information about their vehicle's flight, and some also included displays to show surrounding operations, and/or aspects of the UTM system. For example, Test-Site 3 provided four screens for its GCSO/PIC/USS Op. These individuals were not in line of sight (LOS) of their vehicle. Standard tools shown on their displays were flight planning/execution software, a USS client, and a fusion of radar, multi-lateration systems and GCS telemetry. The fourth screen was available to show other information of the GCSO/PIC/USS Op's choice, including weather, vehicle and USS data, radio frequency usage, etc. Other test-sites, which had more mobile/portable GCSs, used fewer displays. At Test-Site 5, for example, flight crews had a hand-held controller, and one display showing the autopilot software for their vehicle. These flight crews did not have access to a display of UTM information. Instead, UTM information was verbally relayed to them by radio from a centralized location, where the USS Op had such a display.

Table 1. Crew member roles and responsibilities.

Crew role	Crew responsibilities
Pilot-In-Command (PIC)	Served as the main pilot for the vehicle
GCS Operator (GCSO)	Worked the vehicle's flight planning and flight execution software
USS Operator (USS Op)	Monitored and interacted with USS displays (& NASA)
Launch and Software Flight Engineers	Supported specific technical aspects of the vehicle
Visual observers (VO)	Safety monitors who provided visual contact with the vehicles at all times

Table 2. Test site support personnel roles and responsibilities.

Support role	Support responsibilities
UTM manager	Ensured the USS software was running and undertook troubleshooting when needed
Radio control (RC) safety pilots	Served as alternate pilots if the PIC needed assistance
Flight test manager	Coordinated the crews and flights to conduct the test scenarios properly
NASA researchers & observers	Collected observational and survey data. Observers were available to support media day and answer flight team questions

All test-sites used at least one surveillance system to provide information about the airspace not provided by vehicles' on-board sensors (GPS, ADS-B), helping to identify other manned and unmanned aircraft flying near the test-site. A NASA-built iOS application called insight UTM (iUTM), provided visualizations of UTM system information and current operations, and was made available to the test-sites. Test-Sites 3 and 5 elected to use iUTM as an additional situation awareness display. To participate in these tests, participants needed to have certain basic capabilities, but the manner and extent by which the partners met those requirements differed. These differences are not examined in this paper.

2.4 Procedure

During test days, teams of two researchers collected data from the participants at each test-site about their experiences. To the extent possible, researchers observed all flight crews at some point across the test days. Data were collected in a number of ways:

- Observations of participants during flights
- Questionnaires administered at the end of the day
- Interviews conducted at the end of the day

All of these methods solicited feedback on how flight crews used information and where they looked for it. In total, 18 end-of-day group interviews were collected across the six test sites, totaling nearly nine hours of recording. Survey items were presented to the participants in the context of the research objectives of the test-site's test scenarios. Approximately 40 questions were generated in the flight test survey, but conditions were set so that participants only answered around 15 at any one time. There were only three or four questions in each shakedown survey. Most questions used a seven-point rating format, but some were multiple choice or open-ended.

3   Results

Recurring themes that surfaced from field observations convey how information was shared as a function of crew configuration.

3.1   Sources of Information

As the USS interface provides an additional set of information to the crew, team members were asked both before and during flight tests about the source of information they used. Between a shakedown questionnaire and an end of day survey, participants were asked five questions about where they looked during flights to gain information about their vehicle. Numbers of responses were 149 for the end of day survey and were 19 for the shakedown survey.

Before taking part in the flight test, respondents estimated that they would look to other personnel (e.g., the GCSO or the USS Op) three quarters of the time (Fig. 1) and look at displays about 25% of the time to gather information about *their own vehicle*, with talking via radio being the most frequently chosen option (n = 14)[1]. Most respondents reported seeking vehicle information from more than one source. Both integrated and distributed crews report primarily using radio comms (around 20% of the time) and line of sight (17%) to gain information about their vehicle. They did report using tools (e.g., USS client) to gather information, but to a lesser extent than relying on reports from other people. Their responses were very similar for where they would gather information about *other vehicles* flying in the area. Before taking part in the flight test, respondents estimated that they would look to other personnel for information 63% of the time and look at displays about 19% of the time, with talking via radio, again, being the most frequently chosen option (n = 14)[2].

After days when they had flown BVLOS missions, participants were asked a question similar to that above. Most respondents, again, reported looking at more than one source of information to keep track of their vehicle when it was BVLOS. Three sources of information stood out as most frequently used. About 27% of the time, respondents looked at their ground station display and talked to their VO to gather vehicle position information (Fig. 2). The USS client display was the third most popular source of information, chosen 23 out of 135 times (17%). Integrated crews

---

[1] Note that the question options were uneven, as six personnel roles were listed but only three displays.
[2] Note that the question options were uneven, as six personnel roles were listed but only two displays.

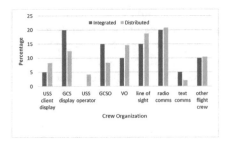

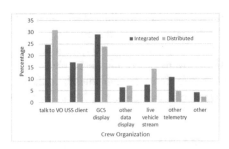

Fig. 1. Percentage of reported usage of information sources by crew configuration before flight test participation.

Fig. 2. Percentage of reported usage of information sources by crew configuration after flight test participation.

reported going to a tool most often to find information about their vehicle, while distributed crews reported they primarily talked to their VO to gain information.

There are two encouraging aspects to the differences between participants' answers before the flight tests and after. The first is that usage of the USS client increased, between shakedowns and post-flight data collection. The percentage of the time participants said they had looked at USS client information more than doubled from just over 7% to just over 17%. The second, is a focusing of where participants looked for information (which is not readily shown in Figs. 1 and 2) from an average of just under four sources (3.8) to fewer than three (2.5).

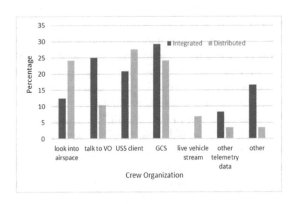

Participants were asked the same question again in the context of acquiring information for altitude-stratified flights. Again, most respondents reported looking at more than one source of information to keep track of relative aircraft positions when their flight was altitude-stratified with another. On average, participants reported looking at three sources of information. Integrated crews looked at their GCS display, the USS client display and talked to their VO most often (Fig. 3). Distributed

Fig. 3. Percentage of reported source usage to obtain information about altitude stratified flights, by crew configuration.

crews looked at their USS client display, their GCS, and into the airspace. In Fig. 3, the slight difference between crews of different organizations, seen in Fig. 2, is more marked, with distributed crews shifting to look at their tools a little more. However, this could be a result of the conditions at the time or the exact nature of the flight.

3.2 Finding and Using Information

When asked questions about the process of finding information rather than the source of data, half of the respondents (16 of 32) reported they were "easily" able to find all the information they needed to support their decisions[3]. Integrated crews seemed to find the information they needed more easily, with two thirds reporting they could "easily find all the information they needed" as opposed to only forty percent of distributed crew personnel.

Ease of finding information plays into how hard crews have to work to achieve their tasks. On average, participants reported their workload as "medium" (m = 4.16 out of 7), with participants from integrated crews feeling they were more loaded (m = 4.8) than participants from distributed crews (m = 3.8) (Fig. 4). It is possible that integrated crews would have a higher load at busier times because the USS Operator was often not an additional person in the team. The USS Op roles had been integrated with other team roles, and were either allocated across the group or were an additional set of tasks for one team member. For distributed crews, the USS Op role was filled by an additional person, who was not located with the team. Under nominal conditions, the crew would not be aware of any additional UTM activity. However, for this series of flight tests, the case where multiple vehicles simultaneously had off nominal events was not tested, and it is hypothesized that distributed crews may experience higher workload or a lower awareness of local operations under these conditions.

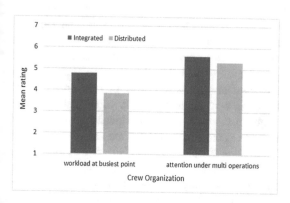

Fig. 4. Mean level of workload and level of attention, shown by crew organization.

Participants were also asked to rate how their level of attention changed when there were multiple vehicles flying. Integrated crew members reported their level of attention changed and they "increased their vigilance" (m = 5.8 out of 7) (Fig. 4). Distributed crew members also increased their vigilance "a little" (m = 5.3), but not as much as integrated crew-members. When there were multivehicle operations, and some of these vehicles were out of the line of sight for the team on the ground, dis-tributed team members may have had less awareness for these other vehicles if they did not make an effort to check in with their USS Op. Respondents often said that they gathered information from other people rather than displays, e.g., "all information was provided by my eyes, crew and OC radio calls" (Test-Site 2 participant). Initially,

---

[3] 9% of respondents reported they had no display, however, absence of a visual display does not mean that the participant received no information, as many flight crews were designed to receive information via voice.

crews intended to look for the potential locations of other vehicles during their flight planning, whereas, while their vehicle was airborne, they intended to only seek out information about other flights if they could see and/or hear this other vehicle in the vicinity. However, participants reported they, in fact, took a slightly different approach, where close to 30% of the time they had looked for other vehicles on their displays "all the time". Integrated crews reported using their displays at some level 75% of the time and distributed crews reported using their displays at some level 38% of the time to gain information about other vehicles. This affected their use of their USS client, a third of the time participants reported looking at the USS displays themselves, and the other two thirds of the time they asked the USS Operator to report the USS client information to them. It should be noted that this depended on a participant's role in the flight crew – whether they were reporting or being reported to. However, discussions highlighted that additional verbal communications were required when the UTM Op was remote from the rest of the crew. A small difference between integrated and distributed crews, that illustrates how the team workflow was different between the two crew organizations, was that 20% of the time, distributed crews reported only gaining USS client information when there was an alert, whereas integrated crews reported that they "always during concurrent operations," or "always," looked at their USS client. This reflects the greater availability of UTM information in integrated crews.

In debriefs, operators were also asked about information that they would want to gain from their displays/tools regarding their own flight and the flights of others. From the debrief transcriptions (see [8] for more information), crews noted they wanted to be able to see immediately all aspects of their vehicle health, performance, and location. They also wanted to be able to find out location and health information about other flights in their vicinity. They were interested in receiving alerts about issues with their own vehicles and with others, and some suggested that they wanted their USS to suggest courses of action, give an account of why issues arise, or how crews might recover from a situation. However, this is a substantial quantity of information, and crews noted occasions during the flight tests when they experienced both visual and aural clutter from their displays. There were concerns that too much data was available and that crews could not pay attention to all of it without being distracted. Although the amount of information that a crew was able to attend to depended to some extent on the size of the team, the debrief and observation comments suggest that information needs to be carefully prioritized and then layered within tools to ensure that the most pertinent information is the most readily available, but all information can be obtained if needed.

Crews were keen to share as much information as possible with support personnel, or "home base," suggesting streaming raw data from their vehicle to these locations. They noted that consistency/standardization of information and formatting on a USS GUI (graphical user interface) is also needed for these remote personnel (including the USS Operator when teams are distributed and s/he is managing a number of flights from a distant location) to be able to compare across, and understand multiple flights. The information that operators felt they should broadcast to others concerned off-nominal vehicle states rather than nominal data. Operators stated that a USS needs to report out, off-nominal events occurring with a user's own operation, such as non-conforming or rogue states, lost links, return-to-base (RTB), and RTB procedures or intent. However, interviewees emphasized that the community needs to agree on the terminology for each of the states that is broadcast. Another point crews made about broadcasting off-nominal states is that the crews themselves cannot be required to send these messages, they will be too busy managing the event. This suggests that off nominal information broadcasted through USS clients will need to be automatically triggered.

4 Discussion

In this paper we reported on data collection efforts from UTM flight tests. Observation data, surveys, and interviews point to at least an informal conclusion that crew configuration guides how UAS flight crews share information. Principle devices for information acquisition about UTM events came from either displays or verbal requests from a UTM operator. Overall, when crews were integrated, they turned to displays for UTM information, while distributed groups preferred verbal communication. Mean rated workload for integrated crews were higher than for distributed crews. This difference may have been due to integrated crews having to burden a single individual with both UTM responsibilities and existing ground station activities. In contrast, the UTM operator was a standalone individual in distributed crews, who did not have a dual role.

To facilitate safe and efficient operations, flight crews needed fast access to easily understandable information about the current mission, nearby operations, and the surrounding environment. As noted above, some feedback suggested it was not a lack of information but that crews sometimes struggled to extract the information they needed from the displays they used in the flight tests – information was buried too deeply in the tool given the time available and other activities occurring, or messages were difficult to interpret. Ease of information access will likely have the most impact on shared SA for integrated crews, based on our results.

Participants reported that there was information that they did not need, although opinion was divided on this. When the information displayed was perceived as unreliable, its usefulness was diminished as operators lost trust and sought out alternate information sources. For example, observers at different sites noted instances of multiple sources for position data showing conflicting information at the same time. Crews considered not just what information they would like to receive but also how it is presented. Many teams liked audio presentation of messages, emphasizing simple wording and that audio presentation should be used selectively. They also noted that the environmental conditions in the field sometimes make visual displays challenging to use. The mode in which information is propagated will likely play the largest role in shared SA for distributed teams. In lieu of the UTM operator being present, verbal communication might have a positive effect on trust and reliance of the information shared – particularly where verbal communication suggests that the shared information has been vetted by another human.

Acknowledgments. The authors would like to thank our partners from academia and industry for their participation in the UTM project.

References

1. Unmanned Aircraft Systems (UAS) Service Demand 2015–2035 Literature Review & Projections of Future Usage, Cambridge, MA (2013)
2. Kopardekar, P., Rios, J., Prevot, T., Johnson, M., Jung, J., Robinson, J.E.I.: Unmanned aircraft system traffic management (UTM) concept of operations. Am. Inst. Aeronaut. Astronaut. (2016)
3. Rios, J., et al.: UTM Data Working Group Demonstration 1 Final Report. Moffett Field, CA (2017)
4. Johnson, M., et al.: Flight test evaluation of a traffic management concept for unmanned aircraft systems in a rural environment. In: Twelfth USA/Europe Air Traffic Management Research and Development Seminar (ATM2017) (2017)
5. Salas, E., Dickinson, T.L., Converse, S., Tannenbaum, S.I.: Toward an understanding of team performance and training. In: Swezey, R.W., Salas, E. (eds.) Teams: Their training and performance, pp. 3–29. Ablex, Norwood (1992)
6. Endsley, M.R., Jones, D.G.: Designing to support SA for multiple and distributed operators. In: Designing for Situation Awareness. 2nd edn., pp. 193–218. Taylor & Francis Group, New York (2012)
7. Wilson, K.A., Guthrie, J.W., Salas, E., Howse, W.R.: Team process. In: Wise, J.A., Hopkin, D.V., Garland, D.J. (eds.) Handbook of Aviation Human Factors, pp. 9-1–9-17. Taylor & Francis Group (2010)
8. Martin, L., Wolter, C., Gomez, A., Mercer, J.: TCL2 National Campaign Human Factors Brief, NASA/TM-2018-219901, NASA Ames Research Center, CA (2018)

Human-Robot Collaborations and Interactions

Intuitive Interfaces for Teleoperation of Continuum Robots

Ryan Scott[✉], Apoorva Kapadia, and Ian Walker

Department of Electrical Engineering,
Clemson University, Clemson, SC 29634, USA
{rscott6, akapadi, iwalker}@clemson.edu

Abstract. This paper presents a novel teleoperation interface for continuous backbone continuum robots. Previous teleoperation interface methods for continuum robots were less intuitive due to a degree-of-freedom mismatch, using non-continuum interface input devices with fewer degrees-of-freedom than the robot that was being operated. The approach introduced in this paper uses a graphical 3D model on screen to directly operate the continuum robot for an easier user experience. This paper details the development of both the model and software. The teleoperation interface was developed specifically for a nine degree-of-freedom pneumatically driven extensible continuum robot, but it can easily be extended to any continuum robot with an arbitrary number of section due to its modular design. Experiments using the aforementioned system on two different continuum robots are reported and areas for future work and improvement are detailed.

Keywords: Continuum robots · User interfaces · Teleoperation

1 Introduction

Robot use has expanded greatly over the last fifty years. The vast majority of robots used in industry are rigid-link manipulators. As the term implies, the links of these robots are rigid and connected at specific joints. These joints are almost exclusively revolute (rotary) or prismatic (linear). Industrial robots are widely used for their ability to perform menial and/or dangerous tasks with high levels of accuracy and repeatability. Their accuracy and repeatability comes at the cost of requiring a structured environment with little, if any, deviation between tasks.

1.1 Continuum Robots

This paper presents a new method of teleoperative control for continuum robots. Contrary to standard rigid-link robots, continuum robots do not have fixed joints and are formed from serially connected smooth "sections". Instead of having rigid links, each section is flexible and a continuum robot can bend at any point along the backbone of each section [1, 2]. This "built-in" flexibility allows for added adaptability when it comes to grasping objects of variable size. This design is naturally inspired by elephant trunks [3, 4] and octopus arms [5]. Small-scale continuum robots have also proven to be very helpful in minimally invasive surgery [6, 7].

The kinematics of a continuum robot are typically based on the constant curvature model, which assumes that each section bends with constant curvature [8]. In this model, each section can be characterized by three parameters – s (arc length), κ (curvature), and φ (orientation) shown in Fig. 1. These three parameters are used as inputs to the kinematic model to determine the position and orientation of the OctArm [5] and Tendril [9] robots (shown to the left and right respectively in Fig. 2) under the new teleoperation approach introduced here.

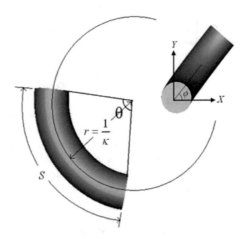

Fig. 1. Continuum robot configuration variables

Fig. 2. OctArm and tendril continuum robots

1.2   Methods of Teleoperation for Continuum Robots

Teleoperation is the process of a human remotely operating a robot. Robots are often used to replace humans in a task that is either too remote, or too dangerous for humans to complete. Teleoperation allows a user to control the robot while remaining out of harm's way for those tasks that still require human supervision [10]. Although there is extensive literature on the teleoperation of rigid-link robots [11], to date very little work has been done on the teleoperation of continuum robots. In the literature, the first teleoperative interface for continuum robots mapped a joystick to a continuum robot using several different mappings [12]. Following this, a human operator manipulated a rigid link robot as an input device to control a continuum robot [13]. Both approaches however, fail to provide a natural mapping since each input device has fewer degrees-of-freedom than the continuum robot being controlled and both are based on kinematically dissimilar input devices to the continuum robot. The degree-of-freedom mismatch complicates control making the approaches [12, 13] difficult to learn as well as making some input motions very difficult to mimic with the robot. This paper presents the first continuum-to-continuum interface for teleoperative control. An intuitive natural mapping is synthesized and implemented, which promises to be easier to use and provides a better user experience.

1.3   Overview

The enabling software follows the client-server model and is thus discussed in two parts: the client and the server. The server renders the user interface and sends data to the client. After receiving robot parameters from the server, the client interprets the data and then sends the appropriate signals to the hardware to move the physical robot. Sections 2 and 3 discuss the server and client software design process, respectively. Section 4 describes the results of several validation experiments run on two robots, the OctArm and the Tendril, to verify the effectiveness of the approach as well as details regarding the hardware of both. The entirety of the code can be found in [14] is also available online at https://github.com/rymasc/x3d-continuum-interface.

2   Server Software Design

The server loads the user interface, runs necessary code to render the graphics model of the robot and sends the appropriate data to the client, which controls the actual robot. Determining the graphics package for the model was the first priority. The selection criteria of the graphics package included use of known programming languages to minimize complexity, ease of use and cost, the ability to mimic bending continuum section motions and ease of integration. In the past, we successfully used VRML [15]. VRML has become mostly defunct [16, 17], and thus its successor, X3D, was chosen.

In X3D, a set of XML tags define a node. Every X3D file has at least one <Scene> node, which contains all the viewpoints, shapes and associated transform nodes. X3D cylinder primitives do not have enough vertices to allow for bending. Instead, we use stacked spheres to approximate the sections. Using this approach, the sphere nodes move individually based on the kinematic model [18] to model the bending and extension needed to accurately depict a continuum robot section. X3D was chosen due to the simple XML syntax and ability to use JavaScript to dynamically modify the sphere nodes.

A transform node wraps each sphere node with a unique id and defines a radius. Forty-five spheres ultimately define each section. The format of the id is *b0-b44*, *m0-m44*, and *t0-44*, for 135 spheres in the robot, where *b* represents the base, *m* the middle and *t* the tip sections respectively. This number can easily be modified depending on the robot being modeled.

2.1 JavaScript Objects

JavaScript allows for the creation of modules, which helps to create repurposeable code. Five object files, or modules define the OctArm model: `Octarm.js`, `Section.js`, `x3dSphere.js`, `Color.js` and `Coordinate.js`. The code has the following hierarchy. `Octarm.js` contains three instances of `Section` for the OctArm base, mid, and tip sections. The `Section` object contains an array of `x3dSphere` objects that each have an associated `Color` and `Coordinate`. Since the model of the robot is a series of spheres, this object oriented programming approach works well for this application.

`Octarm.js` has four instance variables `anchor`, `base`, `mid` and `tip` one member variable n, which represents the number of spheres in each section, and three functions (curve, orient and extend). Each of the sections is formed by making an array variable, `objs[]`, that reads all the appropriate *b0-b44*, *m0-m44* or *t0-t44* tags out of the model. In addition to the instance variable array there is the anchor variable, which corresponds to the last `x3dSphere` in the previous section, or in the case of the base, the anchor itself. There is also a `type` variable and then variables for `radius` and `color`. The `type` variable stores the section type (base, midsection or tip) and the `radius` and `color` variables store the size of the radius and the color (red, green or blue) of the spheres. Finally, there are the robot parameter variables s, k, phi, and d (Defined in Sect. 2.2). `Color` and `Coordinate` are helper files that translate the strings into three member variables. In the case of `Color` they are r, g, b (for red, green blue) and in `Coordinate` x, y, z.

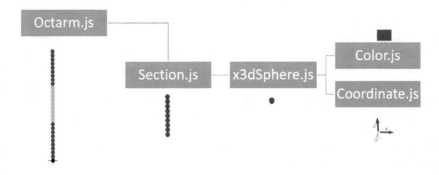

Fig. 3. JavaScript object hierarchy

2.2 User Interface

The 3D model is embedded inside a webpage (a single HTML file), instead of being saved with the usual .x3d extension which allows JavaScript to access the model by selecting elements. The user interface implementation is in `model.html` and contains essentially two components – the controls section and the X3D section. The controls section includes check boxes, button commands and information regarding the parameters of each section of the OctArm. The X3D section places the X3D content in an X3D element, <x3d> </x3d>.

The User Interface (UI) features a feedback segment that allows the user to see the length (s), curvature (κ) and orientation (ϕ) for a given section, or multiple sections. The d parameter mentioned earlier represents the distance between rendered spheres. For more details on the kinematics, see Sect. 2.5 of [14]. It also has check boxes that allow the user to select the parameter they wish to modify for the given section. There is one large row with buttons to call one of the three main OctArm functions - extending, curving, and orienting. These functions increment s, κ and ϕ respectively. Each movement function requires, at least one selected section. The appropriate button increments or decrements that value for the selected section(s), applies it to the model and sends the data to the client program. Note, that it is not valid to select two spatially distinct sections because there is no intuitive reason to move them in unison. The X3D Section contains X3D spheres in the format shown in Fig. 3 but also contains three cylinders and a ViewPoint node all wrapped in Transform nodes. The ViewPoint node ensures the entire model is viewable at an appropriate zoom level. The three cylinders model the X, Y and Z-axes as a triad. Lastly, there is one extra black sphere called the anchor node. The model connects each section to the previous one; tip section to the last sphere in the midsection, which is tied to the last sphere in the base, and the base to the anchor. As far as controlling the view, scrolling the mouse wheel zooms in and out. Holding the left mouse button down and moving the mouse will rotate the view while holding the middle mouse button down and moving the mouse pans. The same action with the right mouse button results in an additional method to zoom in and out (Fig. 4).

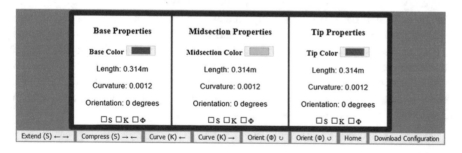

Fig. 4. Control user interface

2.3   JavaScript Libraries

Several JavaScript libraries enable movement of the 3D model in response to the user interface. X3DOM [19] is the most critical library in the project. X3DOM uses the Document Object Model, which is the foundational API for creating dynamic web pages with JavaScript. This library is what enables regular JavaScript code to modify the translation attributes, and anything else in the X3D model, to move the spheres.

X3DOM, developed by the Fraunhofer Institute, is intended to become the official standard for declarative 3D content in HTML5 [19]. There are a few other small libraries in addition to X3DOM and jQuery, as explained in [14].

2.4   Data Transmission

Originally, the X3D JavaScript was intended to run in the web browser, but there is no way to extract the robot parameter data from the browser environment. JavaScript is sandboxed when it runs inside the browser; security exceptions prevent it from opening web sockets and all other forms of file input/output. The *node.js* framework allows JavaScript to bypass these exceptions and directly access the hardware. The model uses three node modules – **ejs**, **express** and **socket-io**. `Server.js` contains the code to create the web server with a designated IP address, render the interface and begin listening for client connections. The socket-io module creates two event handlers – one runs when the server gets a connection and the other runs when the server gets a message. When the server gets a message it sends that message over the port to the Java client code using the `socket.emit()` function – either `OctarmClient.java` or `TendrilClient.java` depending on which robot it is controlling. When the user clicks a button in the UI, it calls the appropriate block of code to modify s, κ, ϕ, move the model, format the data and send it to the server. These parameters are converted to the JSON data structure format. The robot parameters are stored in one key-value pair where the key is `"params"` and the value is an array of nine key-value pairs, to preserve order.

2.5   Server Program Flow

The entry point for this program is an event handler for the web page. Once the document model loads, it immediately creates an instance of `Octarm.js`, which in turn creates 3 instances of `Section.js` and 136 instances of `x3dSphere.js` – 45 of each section plus the anchor node. It does this by creating the robot object, which is an instance of the `Octarm.js` file, in the scope of `Main.js`. The curve, extend and orient functions increment the appropriate values. Afterwards, the transform kinematics function makes all the necessary updates to move the spheres. From there it calls the `fileComm.js` library and writes to the web socket [14].

3   Client Software Design

The client software is responsible for reading a message from the server, interpreting the message and then writing the corresponding values to the hardware to physically move the robot. There are currently two different versions of the client code—one for the OctArm robot and another for the tendril robot; both of these are written in Java. Both are identical in how they receive and parse data; they differ in how they interpret data and then write that interpreted data to hardware. The Java code uses the Socket-IO Java client API allowing for simple communication between JavaScript and Java. The Java program has an event handler that calls a function every time it receives data. The data is received in JSON format and converted into float values. This process is outlined in the Java Socket-IO section. From this point on the code is unique to the robot being controlled.

3.1 Java Socket-IO

The Java code uses Maven to manage the Socket-IO dependency. On startup of the Java code, the `OctarmClient` or `TendrilClient` constructor connects to the socket. Then the constructor defines event handlers for connecting to the socket, disconnecting from the socket, receiving an error from the socket, and receiving a message. The first argument is the event type and the second argument is the function to be called. This function, `call()`, is defined inside the `Emitter.Listener()` object. In both the OctArm program and the tendril program, the first process is to convert the JSON message into an array of floats. This is done by stringifying (a built in JSON Method) the `objects` argument from the `call()` function into a `JSONObject`. Since the JSON Data is in the form of an array, the `JSONArray` is extracted by using the key. From there the float array is built by reading out each element and casting it to a float from a string.

3.2 OctArm Client

The OctArm client receives the s, κ, φ parameters, converts them to voltages and then writes them to the hardware. Simulink was used exclusively to communicate with Quanser DAQ boards [20]. The preexisting Simulink model takes s, κ and φ values as inputs and then converts them into voltages to drive the pressure actuators. Several communication interfaces were evaluated after it was clear that there was no way to directly communicate between JavaScript and Matlab including TCP/IP and C# - ASP. NET. The preexisting Matlab code to convert the configuration variables s, κ, φ could easily be translated into any language so Matlab was abandoned. Java code had already been developed to communicate with the Arduino, which controls the Tendril, so we created a Java to Quanser interface to have a uniform communication system. Quanser has a Java Communications API for Matlab but no direct hardware library. Instead, the Hardware In the Loop (HIL) API was used in combination with the Java Native Interface.

3.3 Tendril Client

The Tendril Client has two classes, `TendrilClient` and `TendrilSystem`. The Socket-IO library is used in the `TendrilClient` class. In addition to the event handler definitions described in Sect. 3.1, there is a function to set up the serial port and a listener class to read from the serial port. This function runs immediately on startup and the listener class `PortReader` is bound to the port. It also instantiates a `TendrilSystem` object which manages the conversion of the s, κ, φ configuration variables into meaningful outputs to control the Tendril. The Tendril hardware is described in more detail in Section IV-C. The Tendril has three sections each controlled by three tendons. These tendons are equally spread around the section 120° apart. Each section's set of motors is offset 40° from the previous one. The base section is controlled by motor 0, 1 and 2 (at 0°, 120°, 240°), the midsection is controlled by motor 3, 4 and 5 (at 40°, 160°, 280°), and the distal section is controlled by motor 6, 7 and 8 (at 80°, 200°, 320°). These values are stored in an array called motorLocations where the index represents the motor number and the value represents its angle. The Tendril is not kinematically modeled as with the OctArm - so unlike the OctArm there was no preexisting code to convert between s, κ, ϕ to output voltages. This conversion process takes place in the TendrilSystem class. All of the s values were left out in this parameterization. The class only uses six parameters to define the robot—the κ and ϕ for each section. Each section's curvature is converted into a percentage of maximum tension by dividing by the maximum curvature for each section, rounding and then converting to a byte. In this model, ϕ values are used to determine which motors to use. The tension found is split between the motors as a function of distance between them, i.e. if ϕ is halfway between motor 0 and motor 1, half of the tension is applied to motor 0 and half to motor 1. Finally, the state of the tension and ϕ values are monitored and if the values change by 15% or more then TendrilClient will write the current configuration to the Arduino. This is done because the Tendril control system is not built to handle rapid continuous changes in the tension value.

3.4 Java Native Interface and C Dynamic Link Library

The Java Native Interface (JNI) allows Java code to call and be called by native code JNI has two prerequisites—a JNI wrapper class and a C dll. The wrapper class consists of two parts. First it needs to load the dll using the System.loadLibrary() method. Second, all functions that need to be called from the dll, must be declared in Java using the native keyword. The C code was written with Visual Studio. There are three main API functions used to create the C dll file: a connect, write and disconnect function. The connection function takes three inputs: the board type, board identifier and a reference to the board variable. The first board has eight analog connections and the second board only has one analog connection, with the combination enabling the nine degrees-of-freedom of the OctArm. The write function has four inputs: the board designation, the channels' designations, the number of channels designated, and the values to write. Function definitions are also different using JNI. The arguments JNIEnv* and jobject must always be included and listed first in JNI. JNIEnv* is a reference to the JNI Environment so that all JNI functions can be used. The second argument, jobject is a Java Object. JNIEXPORT and JNICALL are macros that are not used but are required. In our wrapper class we declared a native function that took type float[] as an input. To include that in the definition we use type jFloatArray}. This pattern persists in JNI for every primitive data type, there is a "j-variant" i.e. jbyte, jchar, jint, jboolean, etc. This is necessary because different compilers use different number of bits for certain data types across languages. This prevents data corruption in primitive data types. JNI Arrays add one more step because of a difference in the way Java stores array variables. In C, arrays simply store the elements in one large block of memory. Java arrays are actually classes and store additional data—such as the size. To use a JNI array it is essential to strip this extra information out to get only the required values. This is done using the JNIEnv* variable.

4 Experimental Validation

The interface described in the previous sections was validated on two continuum robots of different types: the OctArm and the Tendril. In this section, the hardware setup as well as the validation of the interface for each robot is detailed.

4.1 OctArm Hardware

The OctArm is constructed with three continuum sections – a base, midsection and tip, which makes nine controlled variables to define the robot shape: s, κ, ϕ for each section. The OctArm is actuated with pneumatic artificial muscles known as McKibben Actuators. Nine pressure regulators drive the muscles where each section is controlled by three. These pressure regulators are connected to an air compressor and two Quanser Q8-USB control boards. The OctArm also uses nine string encoders to measure the lengths of each muscle group. This research was focused on the OctArm so it was expected that the OctArm would perform better than the Tendril. The preexisting OctArm hardware acts as a black box that takes s, κ, ϕ values as inputs and generates pressure regulator input voltages as outputs. Four validation experiment types were run on the OctArm: each section individually plus one with all sections together. Similar experiments were run on the Tendril to demonstrate the interface could intuitively control the robot.

4.2 OctArm Validation

The purpose of the single section validation experiments was to verify the interface was correctly controlling the robot as well as to measure the accuracy of the system. To properly validate a section, the following script was performed on a given section; extend, compress, curve right, extend, compress, orient clockwise and then return to the starting home position. The full motion videos for the base [21], mid [22], and tip [23] are on YouTube. The interface was quick and easy to use and proved accurate modulo the information in the kinematic model. Gravity pulls the OctArm down, the effect of which is not part of the kinematic model (see Fig. 5 for example). This effect is most evident when moving the base section since the base has the added weight of the midsection and the tip. The effect of not modeling gravity on the tip produced relatively little error because the tip only needs to support itself. However, as the curvature increases these differences become more pronounced (see Fig. 6). The same procedure was run on all sections simultaneously, the results of which are shown in Fig. 7. The same gravitational effects can be observed.

4.3 Tendril Hardware

The Tendril is a thin tendon actuated continuum robot. Its backbone consists of three carbon fiber concentric tubes, with the distal and midsection tubes contracting into each other and the base section tube, respectively. The tubes are spring loaded. Like the OctArm, the Tendril uses three actuators set 120° apart. In this case, the actuators are servomotors. The servo motors connect to an Arduino Mega connected to a PC via

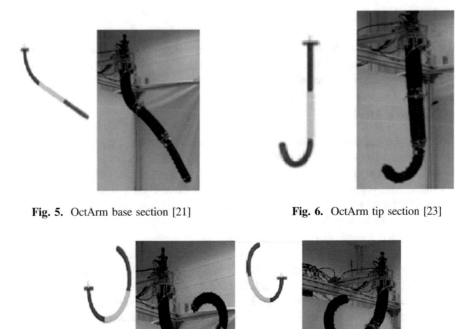

Fig. 5. OctArm base section [21]   **Fig. 6.** OctArm tip section [23]

Fig. 7. OctArm all three sections

serial connection. Unlike the OctArm, which had (via string encoders) feedback of shape, the Tendril's only feedback is via tendon tension sensing (on each of the nine tendons).

4.4 Tendril Validation

Unlike the OctArm, the Tendril control system was not based on a kinematic model using s, κ, φ as inputs. Therefore, a new primitive mapping had to be created to move the Tendril. In this work, κ and φ were used to drive the Tendril, where κ determined the tension a given set of motors would regulate to and φ determined which subset of motors needed to run. Each section's curvature is converted into a percentage of maximum tension by dividing by the maximum curvature for each section of the Tendril. The maximum curvature was estimated by exerting maximum tension on the Tendril and comparing it to the X3D model. Once it matched, the κ value from the feedback section of the user interface was set as the maximum. Then to find the tension to be applied, the percentage of curvature was multiplied by the maximum tension value. This tension value is split between the motors as a function of distance between them. Testing of this primitive mapping was able to show that the Tendril, under user command of the interface developed in this work responded as expected to the interface

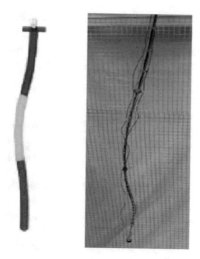

Fig. 8. Tendril S bend [24]

and was able to intuitively move all three sections. The S-Bend shape in Fig. 8 demonstrates this basic capability.

5 Conclusion

Very little work has been done in the field of teleoperation of continuum robots; this work introduces a new concept for intuitive continuum teleoperation. Previous approaches fail to provide a natural mapping. Each input device had fewer degrees-of-freedom than the continuum robot being controlled. This degree-of-freedom mismatch is seen as one of the greatest obstacles to successful teleoperation of continuum robots.

In this paper, for the first time a continuum surrogate software interface was developed to create a natural degree-of-freedom mapping to a continuum robot. Previous controllers were developed, but the operators had kinematically dissimilar interfaces that required extensive practice to gain enough familiarity to operate effectively. The graphical user interface presented herein provides an intuitive natural mapping that is synthesized and implemented using the server client model. This approach is easier to use which provides for a better user experience overall.

Communication with the robot is simplified, because complex conversions are not required prior to issuing commands. The server software loads the graphical user interface, runs necessary code to render the graphics model of the robot and sends the appropriate data to the client. After changes are made in the server software run user interface, the client then reads a message from the server, interpreting the message and then directly writing the corresponding values to the hardware in order to physically move the robot. The user interface provides a low weight, easy to access platform.

This interface uses the new declarative HTML5 standard for 3D graphics, X3D, and was validated with two different continuum robotic systems. Both the Tendril and

OctArm systems were manipulated using this software. These systems were originally designed to be operated by two different software suites; however, the software developed under the approach introduced in this paper was found to be capable of interfacing with both, allowing for easier user control.

References

1. Robinson, G., Davies, J.: Continuum robots – a state of the art. In: Proceedings IEEE International Conference on Robotics and Automation, Detroit, Michigan, pp. 2849–2854 (1999)
2. Walker, I.: Continuous backbone "continuum" robot manipulators: a review. ISRN Robot. **2013**(1), 1–19 (2013)
3. Grzesiak, A., Becker, R., Verl, A.: The bionic handling assistant – a success story of additive manufacturing. Assem. Autom. **31**(4), 329–333 (2011)
4. Hannan, M., Walker, I.: Analysis and experiments with an elephant's trunk robot. Adv. Robot. **15**(8), 847–858 (2001)
5. McMahan, W., Pritts, M., Chitakaran, V., Dienno, D., Grissom, M., Jones, B., Csenscits, M., Rahn, C., Dawson, D., Walker, I.: Field trials and testing of octarm continuum robots. In: Proceedings IEEE International Conference on Robotics and Automation, Orlando, Florida, pp. 2336–2341 (2006)
6. Burgner-Kars, J., Rucker, D., Choset, H.: Continuum robots for medical applications: a survey. IEEE Trans. Rob. **31**(6), 1261–1280 (2015)
7. Trivedi, D., Rahn, C., Kier, W., Walker, I.: Soft robotics: Biological inspiration, state of the art, and future research. Appl. Bion. Biomech. **5**(3), 99–117 (2008)
8. Jones, B., Walker, I.: Kinematics of multisection continuum robots. IEEE Trans. Rob. **22**(1), 43–57 (2006)
9. Wooten, M., Walker, I.: A novel vine-like robot for in-orbit inspection. In: Proceedings 45th International Conference on Environmental Systems, Bellevue, Washington, pp. 1–11 (2015)
10. Stanczyk, B., Buss, M.: Development of a telerobotic system for exploration of hazardous environments. In: Proceedings IEEE/RSJ International Conference on Intelligent Robotic Systems, Sendai, Japan, pp. 2532–2537 (2004)
11. Niemeyer, G., Preusche, C., Hirzigner, G.: Springer Handbook of Robotics, ch. Telerobotics, pp. 1085–1108. Springer, Berlin (2008)
12. Csencsits, M., Jones, B., McMahan, W.: User interfaces for continuum robot arms. In: Proceedings IEEE/RSJ International Conference on Intelligent Robotic Systems, Edmonton, Canada, pp. 3011–3018 (2005)
13. Frazelle, C., Kapadia, A., Fry, K., Walker, I.: Telerobotic mappings from rigid link robots to their extensible continuum counterparts. In: Proceedings IEEE International Conference on Robotics and Automation, Stockholm, Sweden, pp. 1–7 (2016)
14. Scott, R.: Continuum surrogate software interface for teleoperation of continuum robots, Master's thesis, Clemson University (2016)
15. Jones, B., Walker, I.: Three-dimensional modeling and display of continuum robots. In: Proceedings IEEE/RSJ International Conference on Intelligent Robotic Systems, Beijing, China, pp. 5872–5877 (2006)
16. Brutzman, D., Daly, L.: Extensible 3D Graphics for Web Authors. Morgan Kauffman, San Francisco (2007)
17. Web3D Consortium, "What is x3d," (2016). http://www.web3d.org/x3d/what-3d

18. Hannan, M., Walker, I.: Kinematics and the implementation of an elephant trunk manipulator and other continuum style robots. J. Robotic Syst. **10**(2), 45–63 (2003)
19. Fraunhofer-Gesselschaft, "x3dom at a glance," (2016) http://www.x3dom.org/
20. Quanser Inc., "Q8-usb data acquisition device," (2016). http://www.quanser.com/products/q8-usb
21. R. Scott "Octarm base" (2016). https://www.youtube.com/watch?v=dZFECFooogo
22. R. Scott "Octarm midsection" https://www.youtube.com/watch?v=Ru51mc_Gl3o, 2016
23. R. Scott "Octarm tip" (2016). https://www.youtube.com/watch?v=zHtm1-NxOjc
24. R. Scott "Tendril Validation: S bend" (2016). https://www.youtube.com/watch?v=oiLhKiW11XE

Presentation of Autonomy-Generated Plans: Determining Ideal Number and Extent Differ

Kyle Behymer[1($\boxtimes$)], Heath Ruff[1], Gloria Calhoun[2], Jessica Bartik[2], and Elizabeth Frost[3]

[1] Infoscitex, Dayton, OH, USA
{Kyle.Behymer.1.ctr,Heath.Ruff.ctr}@us.af.mil
[2] Air Force Research Laboratory, 711 HPW/RHCI, Dayton, OH, USA
{Gloria.Calhoun,Jessica.Bartik.1}@us.af.mil
[3] Wright State Research Institute, Dayton, OH, USA
Elizabeth.Frost.6.ctr@us.af.mil

Abstract. Autonomous tools that can evaluate a course of action (COA) are being developed to assist military leaders. System designers must determine the most effective method of presenting these COAs to operators. To address this challenge, an experimental testbed was developed in which participants were required to achieve the highest score possible in a specific time window by completing mission tasks. For each task, eight possible COAs were presented. Each COA had four parameters—points, time, fuel, and detection. Four experimental visualizations were evaluated, varying in COA number and type: (1) a single COA (most points), (2) four COAs (four highest point values), (3) four COAs (the most points, the least time, the least fuel, and the least chance of detection), and (4) all eight COAs. Both objective and subjective data indicated that the single COA visualization was significantly less effective than the other visualizations. Suggestions are made for follow-on research.

Keywords: Modeling to generate alternatives · Parallel coordinates plot Plan comparison · Human autonomy interaction

1 Introduction

One of the key components of the military decision-making process is course of action (COA) analysis. Leaders must be able to develop and compare COAs in order to identify the COA that best accomplishes the mission and meets commander's intent [1]. During COA analysis, leaders must consider a wide range of factors including commander's intent, enemy composition, weather and terrain, assets and capabilities, and time available to complete the mission.

Recent advances in autonomous systems have the potential to assist military leaders in analyzing COAs. For example, an intelligent agent has been developed that can enumerate and rank all possible COAs that could potentially fulfill an operator's intent using objective functions [2]. The intelligent agent reasons according to Pareto efficiency, whereby a COA is eliminated from consideration if another COA ranks higher across all relevant parameters [3]. Multiple Pareto optimal COAs may exist within a set

© Springer International Publishing AG, part of Springer Nature 2019
J. Chen (Ed.): AHFE 2018, AISC 784, pp. 90–101, 2019.
https://doi.org/10.1007/978-3-319-94346-6_9

of solutions. Take for example Table 1, which lists the values of six COAs across three parameters (completion time, fuel usage, and the probability of being detected by enemy forces) and italicizes Pareto optimal COAs. COA 1 and COA 2 use the least amount of fuel (10 gallons), but COA 1 takes less time, thus surpassing COA 2. Similarly, COA 3 and COA 4 take the least time (10 min), but COA 3 uses less fuel, thus surpassing COA 4. Finally, both COA 5 and COA 6 have the smallest detection probability, but COA 6 is not Pareto optimal because it takes more time than COA 5. Hence, for the set of COAs represented in Table 1, three COAs (1, 3, and 5) are Pareto optimal. Ultimately, a human decision maker will need to evaluate and accept a single COA from the set of Pareto optimal COAs [2].

Table 1. Pareto optimal COAs italicized.

COA	Time	Fuel	Detection
1	*15 min*	*10 gal*	*50%*
2	22 min	10 gal	50%
3	*10 min*	*20 gal*	*30%*
4	10 min	25 gal	30%
5	*30 min*	*40 gal*	*10%*
6	35 min	40 gal	10%

If an operator only needs to consider three parameters, perhaps Table 1 is a "good enough" representation of the Pareto space. However, as the number of parameters increase, visualizing the Pareto space becomes more challenging making it more difficult for the human decision maker to select a single COA [3]. Operators need a visualization method that allows them to compare N COAs across N parameters. Recent research has suggested that a parallel coordinates plot can provide this capability [3]. The parallel coordinates plot uses a series of parallel axes to show how multiple COAs vary across multiple parameters. For example, Fig. 1 illustrates how eight different COAs (A through H) vary across four criteria (quality, time, fuel, and detection). The tradeoffs between COAs are immediately visible. For example, COA G (colored orange) has the highest quality but consumes the most fuel. COA B (colored mustard) consumes the second smallest amount of fuel but takes the most time.

A significant challenge facing the design team is determining the ideal number of alternative COAs to present to the operator as each one imposes information retrieval costs. Moreover, presenting alternatives can potentially mask aspects of the problem space and influence the operator's cognitive processing (e.g., inducing a perceptual framing effect) [4]. One approach is to present a single solution to the operator, the solution that the autonomy has determined to be the best. Another approach involves modeling to generate alternatives (MGA), focusing on generating a small set of alternatives that are "good" in terms of achieving the operator's goal but different in respect to the relevant parameters of the solution space [5]. This approach aims to generate options that can achieve the commander's intent but vary in other parameters. For example, in Fig. 1, COAs A, B, G, and H are all relatively high in quality, but

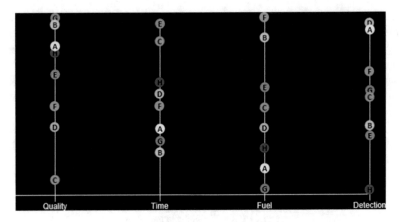

Fig. 1. Parallel coordinates plot representing eight COAs across four parameters. For each parameter, higher is better (e.g., COA E will use the least time, COA F will use the least fuel, COA D has the smallest chance of being detected).

COA B takes the most amount of time, COA G uses the most amount of fuel, and COA H has the highest chance of being detected.

To address this challenge, an experimental testbed was developed in which participants were required to achieve the highest score possible in a specific time window by completing mission tasks with eight vehicles. Eight possible COAs were provided for each task (one for each vehicle). Each COA had four associated parameters—quality points, time, fuel, and detection. For example, using vehicle A to complete Task 1 might earn 25 points, take 16 min, cost 13 gallons of fuel, and have a 50% chance of being detected, while using vehicle B might earn 10 points, take 5 min, cost 12 gallons of fuel, and have a 17% chance of being detected. Four visualizations were evaluated, varying in number of COAs represented and type: (1) a single COA (most points), (2) the four COAs with the four highest point values, (3) the four COAs with the highest value for each parameter—the most points, the least time, the least fuel cost, and the least chance of detection, and (4) all eight COAs.

Figure 2 illustrates how the same data set was depicted in each of the four visualizations. In (a), a single COA is shown (COA C) that will earn the most points (i.e., highest quality). In (b), four COAs are shown (COAs C, F, B, and E) that have the four highest possible point totals. In (c), four COAs are shown (COAs C, G, H, and A) that have the highest value for each parameter respectively—COA C has the most points, COA G takes the least amount of time, COA H uses the least fuel, COA A has the smallest chance of being detected by enemy forces. Finally, (d) displays all eight COAs.

No matter the visualization, participants had the ability to 'drill down' to see the values for all eight COAs to select the one they thought best. Two drill down methods were available (see Fig. 3). In the first, participants could mouse over a parameter in the parallel coordinates plot to see the values of each of the eight COAs for that parameter. Figure 3 depicts the results of a mouse over of the quality parameter. In the second method, participants could mouse over a vehicle button to see that vehicle's values for

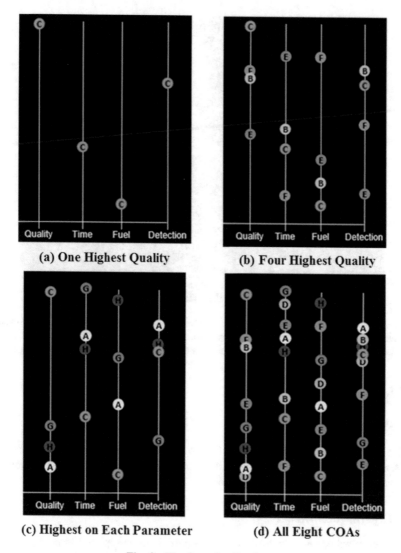

(a) One Highest Quality

(b) Four Highest Quality

(c) Highest on Each Parameter

(d) All Eight COAs

Fig. 2. The four visualizations.

each of the four parameters. Figure 3 depicts the results of a mouse over of vehicle F, and shows that selecting vehicle F will result in 21 points, take 32 min, use 16 gallons of fuel, and has a 5% chance of being detected.

The present study examined which visualization was most effective at aiding participant performance. Both objective (e.g., response time, score) and subjective (e.g., perceived workload, perceived performance) measures were collected. The gaming-type experimental testbed developed to manipulate the type and number of COAs is also described, as this approach can support further research examining human-autonomy teaming in decision making.

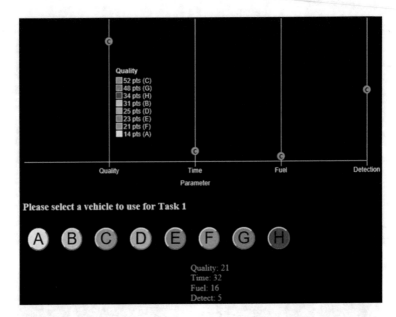

Fig. 3. Two drill down methods. Mouse overs of a parameter (e.g., Quality) showed the parameter's specific value for all eight COAs, from highest to lowest. Mouse overs of a vehicle selection button (e.g., F) showed the specific values of all four parameters for that specific COA.

2 Method

2.1 Participants

Twelve volunteer employees working at a U.S. Air Force Base between the ages of 25–57 (M = 33.5, SD = 9.33) participated in the study. All participants reported normal or corrected normal vision and color vision.

2.2 Experimental Design

Trials were blocked by visualization (e.g., a participant completed trials with the One Highest Quality COA displayed, then trials with the Four Highest Quality COAs displayed, then trials with the Four Highest on Each Parameter COAs displayed, and finally trials with All Eight COAs displayed) with the order of the four visualizations counterbalanced across participants. Within each of the four blocks, participants completed five trials. For each trial participants were presented a series of tasks and were trained to select one of the eight COAs (A-H) to complete each task. Each COA differed on four parameters: quality (i.e., the number of points the participant could accumulate for completing the task with the vehicle associated with the COA, time to complete the task, fuel used, and probability of detection).

2.3   Test Stimuli

A range of possible values was determined for each factor: Quality 10–60 points, Time 15–45 min, Fuel 5–22 gallons, and detection probability 5%–60%. The total simulation mission time of 480 min was divided by the fastest possible task completion time (15 min) resulting in the need to generate 32 sets of 8 values (see example set in Table 2). The following procedure generated this data. First, a random number generator was used to create 32 sets of 8 values for each factor, within the acceptable range for each factor. In order to balance each set of eight COAs, Table 3 was used to sort the randomized values into COAs that had tradeoffs. For example, COA 1 was assigned the 4th highest randomized quality value, the 3rd quickest randomized time value, the 6th most fuel consumed value, and the highest probability of detection. For each of the 32 sets, each of the 8 COAs was randomly assigned to a vehicle letter. Thus, for Task 1, vehicle A might correspond to COA 1 in Table 3, but for Task 2, vehicle A might correspond to COA 7 in Table 3. The colors used to differentiate the COAs were selected per Post and Goode's (2017) recommendations [6].

Table 2. Example set of COA values across the four parameters for a given task.

COA	Quality	Time	Fuel	Detection
A	36	16 min	11 gal	53%
B	43	22 min	17 gal	60%
C	27	33 min	08 gal	33%
D	49	40 min	19 gal	16%
E	58	45 min	10 gal	48%
F	12	31 min	15 gal	5%
G	10	18 min	13 gal	37%
H	60	42 min	22 gal	35%

2.4   Trial Procedure

To begin a trial, a participant clicked a button labeled *BEGIN*. Participants then began the task depicted in Fig. 4. Participants could use the parallel coordinates plot (labeled *Vehicle Comparison Tool* in Fig. 4) to determine which vehicle to select for a task. Once the participant's decision was made, he or she clicked the associated vehicle button (labeled *Vehicle Selection Buttons* in Fig. 4).

Upon COA/vehicle selection, the *Scoreboard* updated to show the results of the participant's selection. *Score* increased by that vehicle's associated quality value if the vehicle wasn't detected. *Score* remained the same and the *Detections* increased by one if the vehicle was detected. *Time Remaining* decreased based on the amount of time the selected vehicle needed to accomplish the task. The *Fuel Status Display* (see Fig. 5) also updated, reducing the fuel level of the selected vehicle by the amount of fuel used to complete the task. If any vehicle had less than 50 gallons of fuel, *Fuel Violations* increased by one. At this point, participants received the next task. Participants

Table 3. COAs balanced across the four parameters.

COA	Quality	Time	Fuel	Detection
1	4	3	6	8
2	5	1	3	7
3	3	6	7	2
4	6	5	1	3
5	7	4	5	1
6	8	2	4	5
7	1	7	8	4
8	2	8	2	6

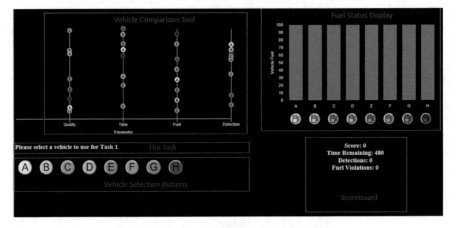

Fig. 4. Screenshot of the experimental testbed.

continued to receive new tasks until no vehicles could complete a task in the remaining mission time, at which point the trial ended.

2.5 Test Sessions

Upon arrival, participants read and signed the informed consent document and completed a demographics questionnaire. Next, participants were trained on the experimental tasks, beginning with a discussion of the four parameters they needed to consider when selecting a COA. Participants learned they had four equally important goals: (1) maximize points, (2) maximize the number of tasks completed, (3) minimize fuel violations, and (4) minimize detections by enemy forces. Participants were then briefed on the five major display components depicted in Fig. 4. Finally, participants were trained on the four visualizations and both drill down methods.

After each block of trials, participants were given a post-block questionnaire asking about their performance, workload, ability to identify the best COA, ability to identify the best COA for a specific parameter, how often they drilled down, and the strategy they used for the visualization they just experienced. All questions used a five point

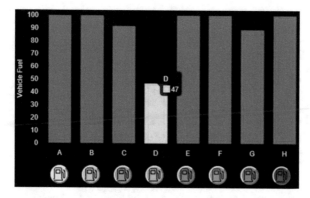

Fig. 5. Fuel Status Display. If a vehicle had 50 or less gallons of fuel its fuel bar turned yellow. A mouse over of a vehicle's fuel bar displayed the number of gallons of fuel remaining (e.g., see vehicle D's fuel bar). Clicking the corresponding refuel button located beneath each vehicle's fuel bar refueled the vehicle to 100%. Each refuel cost 25 min of mission time.

Likert scale, with the exception of the strategy question, which was open-ended. After trials with all four visualizations were finished, participants completed a final questionnaire asking them to rank the four visualizations in terms of their performance, speed, accuracy, workload, and frequency of drill down. The final questionnaire also contained several open-ended questions asking participants which visualization they most and least preferred, whether or not their strategy changed depending on the visualization, which drill down method they preferred, and for display improvement suggestions. Total session time, per participant, was approximately 2 h.

3 Results

Performance data for each participant were collapsed across the five trials for each visualization. Objective data were analyzed with a repeated measures Analysis of Variance (ANOVA) model. An ANOVA model was also applied to the post-block questionnaire responses. The final questionnaire ranking data were analyzed using the Friedman nonparametric test of significance. Post-hoc Bonferroni-adjusted t-tests were performed for significant ANOVA and Friedman results.

3.1 Objective Results

The results of an ANOVA indicated that the number of completed tasks per trial significantly differed across COA visualizations ($F(3, 33) = 7.33$, $p = .003$, $\eta_p^2 = .40$). Participants completed significantly more tasks in the Highest on Each Parameter visualization as compared to both the One Highest Quality ($t(11) = 3.16$, $p = .05$, $d = 1.08$) and Four Highest Quality ($t(11) = 3.26$, $p = .05$, $d = 1.11$) visualizations (see Fig. 6). There were no significant effects with respect to other performance data, including score, response time, detections, and fuel violations.

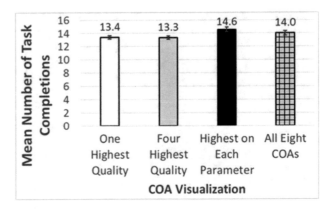

Fig. 6. Mean number of task completions for the four COA visualizations. Error bars are standard errors of the means.

3.2 Subjective Results

Post-Block Questionnaire. ANOVA results indicated that perceived workload was significantly different across COA visualizations, with participants rating their workload significantly lower in the All Eight COAs visualization as compared to the One Highest Quality visualization. An ANOVA also indicated that participants' perceived ability to determine which COA was best for a particular parameter significantly differed across visualizations; participants rated the Highest on Each Parameter and All Eight COAs visualizations significantly better than the One Highest Quality visualization. ANOVA results showed participant ratings of drill down frequency significantly differed across visualizations, with the One Highest Quality visualization rated as requiring more drill down occurrences than the All Eight COAs visualization. Refer to Table 4 for ANOVA results and Table 5 for post-hoc results.

Table 4. Post-block questionnaire ANOVA results.

Variable name	$F(3, 33)$	p	η_p^2
Workload	6.02	.002	.35
Best parameter	8.82	.0002	.44
Drill down	6.02	.002	.35

Final Questionnaire. Results of a Friedman test indicated significant differences in how participants ranked their performance, speed, accuracy, workload, and drill down frequency (see Table 6; post-hoc results are shown in Table 7). Participants ranked their performance as significantly better and their speed as significantly faster in the All Eight COAs visualization as compared to the One Highest Quality visualization. Participants also ranked their accuracy as significantly higher and their workload as significantly lower in the All Eight COAs and Highest on Each Parameter

Table 5. Post-block questionnaire post-hoc results.

Variable name	One highest quality M	SE	Highest on each parameter M	SE	All eight COAs M	SE	t(11)	p	d
Workload	3.92	0.26			2.58	0.32	3.75	.02	1.34
Best parameter	2.83	0.30	4.67	0.26			4.02	.01	1.72
Best parameter	2.83	0.30			4.67	0.26	4.16	.01	1.91
Drill down	3.92	0.26			2.58	0.31	3.75	.02	1.34

visualizations as compared to the One Highest Quality visualization. Although the results of a Freidman test indicated a significant rank order of perceived drill down frequency, no post-hoc comparisons were significant.

Table 6. Final questionnaire Friedman results.

Variable name	$\chi^2(3)$	p	W
Performance	8.70	.03	.85
Speed	11.50	.01	.98
Accuracy	11.50	.01	.98
Workload	11.70	.01	.99
Drill Down	8.40	.04	.84

Seven participants indicated they most preferred the All Eight COAs visualization, four preferred the Four Highest Quality visualization, and one preferred the Highest on Each Parameter visualization. Eleven out of twelve participants reported that the One Highest Quality visualization was their least preferred visualization, typically citing this visualization's lack of information compared to the other visualizations. The remaining participant least preferred the All Eight COAs visualization because "too many options were shown".

4 Discussion

Results indicated that both objective and subjective data were definitely least favorable for the One Highest Quality visualization compared to the other three visualizations. The majority of participants also indicated that the One Highest Quality visualization was the least preferred visualization. This result is unsurprising as the One Highest Quality visualization provided less immediately available information about the eight COAs to participants.

However, the results did not indicate a clear best option from the other three visualizations. Participants completed significantly more tasks in the Highest on Each Parameter visualization as compared to both the One Highest Quality and Four Highest Quality visualizations. This result may have been because the COA that took the least

Table 7. Final questionnaire post-hoc results.

Variable name	One highest quality		Highest on each parameter		All eight COAs		$t(11)$	p	d
	M	SE	M	SE	M	SE			
Performance	3.42	0.26			2.00	0.25	7.34	.0001	1.62
Speed	3.58	0.29			2.25	0.22	4.69	.004	1.51
Accuracy	3.58	0.23	2.00	0.21			3.98	.01	2.07
Accuracy	3.58	0.23			2.17	0.28	7.34	.0001	1.63
Workload	3.58	0.23	2.33	0.19			3.80	.02	1.72
Workload	3.58	0.23			2.00	0.25	6.92	.0002	1.92

amount of time was easily determined in the Highest on Each Parameter visualization, thus enabling participants to more rapidly evaluate time, allowing them to complete more tasks. The subjective data statistical results did not significantly differ between the Highest for Each Parameter, Four Highest Quality, and All Eight COAs visualizations. A slight majority of participants (seven out of twelve) did prefer the All Eight COAs visualization.

One limitation of the study was that participants had as much time as they desired to drill down to additional information about COAs that were not represented on the parallel coordinates plot. This could have masked the benefit of the parallel coordinates plot as well as any differences between the visualizations. Future research should consider temporal demands for task completion, limiting the time available for the participant to drill down. The incorporation of additional tasks is another possibility, creating a more realistic control station environment where an operator needs to manage attention effectively across multiple tasks.

An additional limitation may have been the lack of a "best COA" for each task. The methodology which systematically balanced the eight COAs across the four parameters also made it more ambiguous as to which COA was the best option for a specific task. This may have minimized performance differences across visualizations. Thus, a more complex mission environment would need to be developed that results in varying acceptability of different COAs. To accomplish this, such research would benefit from including an intelligent agent that can reason about the experimental task and provide participants with one or more recommendations. For example, in one visualization an intelligent agent could recommend a single COA based on a weighting across the four parameters. Another visualization could use the MGA approach [5] with the intelligent agent returning solutions that meet desired criteria (e.g., a minimum number of points) but differ across the remaining parameters. For follow-on human-agent teaming research, it is recommended that this gaming-type experimental testbed be enhanced and utilized as the task environment is engaging and rapidly trained.

Acknowledgments. This work was funded by the Air Force Research Laboratory.

References

1. Center for Army Lessons Learned: MDMP: Lessons and Best Practices. No. 15-06 (2015)
2. Hansen, M., Calhoun, G., Douglass, S., Evans, D.: Courses of action display for multi-unmanned vehicle control: a multi-disciplinary approach. In: The 2016 AAAI Fall Symposium Series: Cross-Disciplinary Challenges for Autonomous Systems, Technical Report FS-16-03 (2016)
3. Behymer, K.J., Mersch, E.M., Ruff, H.A., Calhoun, G.L., Spriggs, S.E.: Unmanned vehicle plan comparison visualizations for effective human-autonomy teaming. In: 6th International Conference on Applied Human Factors and Ergonomics (2015)
4. Smith, P.J.: Making brittle technologies useful. In: Smith, P.J., Hoffman, R.R. (eds.) Cognitive Systems Engineering: The Future for a Changing World, Chap. 10, pp. 181–208. CRC Press, New York (2018)
5. Brill, E., Flach, J., Hopkins, L., Ranjithan, S.: MGA: a decision support system for complex, incompletely defined problems. IEEE Trans. Syst. Man Cybern. **20**(4), 745–757 (1990)
6. Post, D.L., Goode, W.E.: A color-code design tool. In: 19th International Symposium on Aviation Psychology (2017)

Trust in Human-Autonomy Teaming: A Review of Trust Research from the US Army Research Laboratory Robotics Collaborative Technology Alliance

Kristin E. Schaefer[1(✉)], Susan G. Hill[1], and Florian G. Jentsch[2]

[1] United States Army Research Laboratory,
Aberdeen Proving Ground, Aberdeen, MD, USA
{kristin.e.schaefer-lay.civ,
susan.g.hill.civ}@mail.mil
[2] University of Central Florida, Orlando, FL, USA
florian.jentsch@ucf.edu

Abstract. Trust is paramount to the development of effective human-robot teaming. It becomes even more important as robotic systems evolve to make both independent and interdependent decisions in high-risk, dynamic environments. Yet, despite decades of research looking at trust in human-interpersonal teams, human-animal teams, and human-automation interaction, there are still a number of critical research gaps related to human-robot trust. The US Army Research Laboratory Robotics Collaborative Technology Alliance (RCTA) is a 10-year program with government, industry and academia combining to conduct collaborative research across four major robotic technical areas of intelligence, perception, human-robot interaction, and manipulation and mobility. This paper describes findings from over 60 publications and 49 presentations describing research conducted as part of the RCTA from 2010 to 2017 to address these critical gaps on human-robot trust.

Keywords: Human-robot interaction · Teaming · Trust

1 Introduction

Autonomous systems, both embodied robots and non-embodied artificial intelligence (AI) systems in software, will be important in future military operations. Such systems have the potential to provide a decisive edge in future conflict. When the US Army Research Laboratory (ARL) Robotics Collaborative Technology Alliance (RCTA) was first starting in 2010, the vision for the future was, and continues to be today, effective manned-unmanned teaming. For Army operations, this means having Soldiers interact with autonomous unmanned systems much as they would interact with fellow Soldiers. Examining issues of how humans interact with robots is critical to that vision. The field of human-robot interaction (HRI) is an interdisciplinary field that seeks to understand, design, and evaluate robotic systems used by humans [1]. It focuses on the interactions between human users and systems, including the user interface and the underlying

© Springer International Publishing AG, part of Springer Nature (outside the USA) 2019
J. Chen (Ed.): AHFE 2018, AISC 784, pp. 102–114, 2019.
https://doi.org/10.1007/978-3-319-94346-6_10

processes that produce the interactions. As autonomous systems technology advances, so too does the interest in HRI. Specifically, the roles, such as supervisor, peer, bystander [2], and the nature of these interactions within these roles, needs to be better understood.

Collaborative technology and research alliances (CTAs and CRAs) are partnerships used by ARL to establish integrated and coordinated programs of basic and applied research by a consortia of industry and academic partners with government organizations, to work towards an identified vision and specific research goals[1]. The stated purpose of the RCTA in the 2017–2018 Biennial Program Plan[2] was to "bring together government, industrial, and academic institutions to address research required to enable the deployment of future military unmanned ground vehicle systems ranging in size from man-portables to ground combat vehicles". The four key technology areas expected to be critical to the development of autonomous systems were perception, intelligence, HRI, and dexterous manipulation and unique mobility. It was apparent that these four technical areas were not fully independent – there are much overlap and many interconnections among them.

We were involved specifically with the HRI technical area, and the interconnections of the human element with the other technology development-focused research areas. The focus on developing human-robot teams, and knowledge of gaps in HRI research led to development of thrust areas that focused on communications between humans and robots, elements of teaming, and the social and cultural implications for HRI and robotic technology development. More specifically, the topic of trust has been a critical component of HRI since the RCTA began. Trust has been identified frequently as an issue in meetings with military leaders as it was generally recognized that lack of trust in military operations can have a significant negative impact on mission performance. This was particularly thought true as the concepts of autonomous robotic systems interacting with Soldiers as a team were developed and researched.

2 Human-Robot Trust

Many people have tried to develop an operational definition of trust, especially as it applies to human-robot interaction. However, as Adams and colleagues [3] stated, there is still no consensus on a definition today. Consequently, one of our first efforts within the RCTA was to better classify trust as an operational construct. In a review of the interpersonal, automation, and human-robot trust literature, over 300 definitions of trust were identified that underscored the importance of expectations, confidence, risk or uncertainty, reliance, and vulnerability as core concepts involving trust [4, 5]. However, review of the experimental research suggested that there was more to the development of trust than just the aforementioned considerations. In part, the difficulty in developing an operational definition of trust stems from the fact that there are

---

[1] More information about ARL's past and current CTAs and CRAs are located on ARL's website http://www.arl.army.mil/www/default.cfm?page=93.

[2] The biannual program plan is available at http://www.arl.army.mil/www/default.cfm?page=392.

different types of trust-related constructs. These can be broken down into four main types: (a) *trust propensity* is a stable trait unique to the individual; (b) *trustworthiness* is developed from cursory information interpreted by the individual about the characteristics of the robot partner; (c) *affective-based trust* is an emergent attitudinal state in which the individual makes attributions about the motives of the robot partner; and (d) *cognitive-based trust* can change based on the interactions with the robot over time [5]. Further, there are different phases to the trust process, including trust development, trust calibration, and trust-based outcomes such as reliance, compliance, complacency, and general use (see Fig. 1). An example of this trust process was provided by Phillips et al. [6] who described a relationship between subjective trust and the trust-based outcome to use or not use a robot during a foot pursuit. Low trust led to disuse of the robot, resulting in the requirement of sending a Soldier to complete the task.

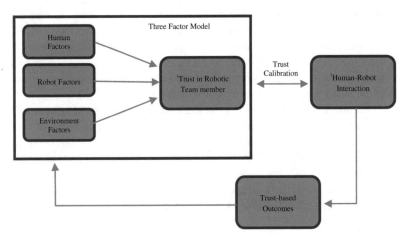

Fig. 1. Trust model including subjective[1], behavioral, and physiological[2] trust measurement (adapted from [7])

To understand the trust process, researchers in the RCTA conducted a full review of the extant literature. Then, meta-analyses of human-robot trust and human-automation trust confirmed the development of a "Three Factor Model of Human-Robot Trust" [8, 9] as depicted in the boxed-in portion of Fig. 1. This work identified antecedents of trust development associated with the human (e.g., traits, states, cognitive factors, and affective factors), robot (e.g., features and capabilities), and environment (e.g., task-related and team specific factors). Moreover, it highlighted critical gaps related to HRI research: intra-team behavior (e.g., team collaboration including mental models and shared situation awareness), bidirectional communication (e.g., the need for robot communication concerning errors, feedback, and intent; and transparency), and social interaction (e.g., social cognition). It is within these three areas that the RCTA researchers have made the most progress in advancing the field of human-robot teaming.

3 Intra-team Behaviors

Optimized intra-team behaviors enable the teaming of intelligent agents with Soldiers. This promises to expand the small unit sphere of influence to enable a greater level of autonomy for heterogeneous systems and support advanced operations. However, the addition of an intelligent agent, such as a robot, can change the trained interpersonal team dynamics. A top priority becomes the development of common ground to understand the task and team roles and capabilities. This is essential because any deviation between expectations (on the part of the human team members or the robot) and actual behavior can lead to a degradation in trust. Therefore, the effectiveness of the intra-team behaviors that support trust development is dependent on developing appropriate mental models and maintaining shared situation awareness.

3.1 Mental Models

Mental models are internal representations that can support prediction, explanation, or add additional understanding to potential interactions with objects, people, or across tasks [10]. Mental models have a long history of being researched in human and human-agent teams. Whereas accurate mental models can enhance teaming operations, incomplete or inaccurate mental models can lead to disastrous consequences, as those seen in human team performance (e.g., USS Vincennes incident, Three Mile Island accident). Therefore, a specific goal of the RCTA was to determine the role of mental models on human-robot teams, possible means to measure or quantify mental models, and the capability to develop shared mental models between human and robot team members.

In a review of literature on human-robot teams, mental models of a robot's intelligence and collaborative capabilities are influenced by a robot's appearance and communication capabilities, as well as the person's features (e.g., personality traits and culture) [11]. These factors may prompt people to form inaccurate or overly presumptuous mental models of the robot's functional capability or knowledge base, thereby negatively affecting collaborative tasking and causing degradations in trust, which subsequently results in disuse or misuse of the robot. This may be due in part to the fact that mental models are often formed around an individual's prior experience with a system or similar system [12]. However, most people have limited prior experiences with robots, and thus rely on initial perceptions based on physical form. Additional RCTA research has supported this notion through several studies that showed that the physical form of robots (across robot domains) directly influenced ratings of perceived intelligence and mental models classifying a machine as a robot. For example, more anthropomorphic-appearing robots were perceived to be more intelligent and more robot-like. Further, these perceptions were found to predict initial trustworthiness (trust prior to an interaction) of a robot [13, 14]. Physical features of humanoid robots can influence feelings of likeability (e.g., gender-neutral robots are more likeable), fear (e.g., human-like robots are more feared), and can influence trust [15]. Overall, incomplete mental models can be easily influenced by media, physical form, reputation, and prior interactions [11, 13, 15, 16], all of which may lead to an inaccurate judgment of the robot's actual capabilities.

To better understand the actual implications of mental models on trust calibration, a second issue explored by the RCTA was how to quantify and index human mental models of robots. One technique was to have people draw a picture of an ideal robotic military teammate to assess early mental models [17]. The drawing was then evaluated through the classification of the presence or absence of a weapon, anthropomorphism, zoomorphism, and methods of vehicle locomotion (e.g., wheeled, legged, tracked, or other). Outcomes of these assessments were compared to participants' answers to questions assessing knowledge about technology capabilities, task, and team interaction. A key finding from this methodology was that both anthropomorphism and zoomorphism in the drawings were related to understanding of capabilities and performance. A second approach was developed to objectively quantify mental models of humans and algorithms of spatial decision-making [18]. A divergence value between all the human and algorithm solutions was computed using the Algorithm for Least Cost Areal-Mapping between Paths (ALCAMP). Then, the Kruskal's Non-metric Multidimensional Scaling (MDS) algorithm was used to visually represent similarity space of solutions, and depict the likelihood of a person adopting another team member's solution. By assessing the decisions made during this spatial planning task, it was possible to infer the strategies that were adopted by both the human and algorithm. This can be used to support a shared understanding and quantify the potential for shared mental models.

Shared mental models are "knowledge structures held by members of a team that enable them to form accurate explanations and expectations for the task, and, in turn, to coordinate their actions and adapt their behavior to the demands of the task and other team members" [19]. They help with prediction of team member actions, support adapting decisions based on team demands, and coordinate actions especially in circumstances where there is limited communication capability [20]. For human-robot teams, there remain more questions than answers on how to develop effective shared mental models to develop appropriate trust. To start to bridge this gap, Ososky and colleagues [21] describe a number of these barriers for developing shared mental models including the structure for collaborative transfer of data and knowledge, identification of relevant information, and how team members should use this knowledge to act upon the world.

3.2   Situation Awareness

Situation awareness (SA) is defined as "the perception of the elements in the environment within a volume of time and space, comprehension of their meaning, and the projection of their status in the near future" [22]. It is important for human-robot teaming for two main reasons. First, changes in information presentation, human vigilance and engagement, and current level of trust can all influence a human team member's SA of the robot and of the task [23]. Second, the development of autonomy that operates with an accurate situation model (robot SA) is needed to support effective human-machine teaming, coordinate teams of multiple machines, operate in complex contested environments, and support safe operations in unanticipated, dynamic environments [24].

A main goal of the RCTA research was to lay the foundations that enable robots to support multi-state missions without a specific location as an endpoint. To reach that goal, teams had to operate at realistic operational tempos in unstructured, unfamiliar environments, where accurate and cohesive SA was critical to success. One study investigating how reliability, or rather unreliability, of operator aids affected SA, found that reliable diagnostic aiding is perceived as useful and supports SA development [25]. Perhaps more interesting was the finding that the effect of an unreliable diagnostic aid on SA was dependent on the amount of information available to the operator.

A second thrust was to look at developing shared SA between human-robot teams, whereby shared SA is "the degree to which team members possess the same SA on shared SA tasks" [26]. The RCTA found that shared SA is difficult because a robot's cognitive architecture or world model may not represent the world or process the mission goals in the same way as human team members. More specifically, AI algorithms are inherently bad at processing and reasoning about contextual information [27, 28]. This often leads to different or unexpected decision-making during teaming operations. In our research looking at path planning, we found that there are often multiple "human ways" of making decisions that are different from decisions algorithms make [18, 29]. This is in part due to individual contextual understanding and use of different strategies for decision-making. This discord between human and robot decision-making is where trust has the opportunity to degrade. Therefore, adding in processes for developing bidirectional communication to better communicate intent and provide reasoning explanations can help support development of shared SA.

4 Developing Bidirectional Communication Paradigms

Bidirectional communications can help develop trust and SA. Information will be exchanged through both implicit (e.g., changes in behavior) and explicit (e.g., natural language) modalities. An important benefit of bidirectional communication is the capability to provide reasoning to specific decision-making processes. Requests for additional information and clarifications can enhance teaming and effective performance.

4.1 Communicating Intent

Within the confines of future human-robot teams, individual decisions made by both human and robot team members are likely to dynamically influence the actions or decisions of the rest of the team [30]. Since an intention to perform an action is a predictor of performance [31], the need to clearly communicate these intentions between team members is essential. A goal of the RCTA was to understand the benefits, limitations, and challenges associated with human-to-robot and robot-to-human intent communication, and determine appropriate types of equipment to facilitate the communication and engender trust in the team [32, 33]. There are a number of ways to communicate intent: changes in behavior, natural language, gaze, and gestures, amongst others. But perhaps the most important factor to communicating intent was that people and algorithms do not communicate or interpret transmission of the same information in the same way. For example, drivers on public roadways will often use signals or gestures with driverless vehicles that are considered to be socially accepted means for communicating both their intentions to other drivers, as well as requests for other driver responses (e.g., turn signals, or waving another driver forward). While these are socially-driven and accepted rules, a driverless vehicle may be unable to interpret all of these modalities of communication. Sometimes it is only possible for a robot to communicate through changes in behavior (e.g., stop rate, rate of speed change, etc.), and these behavioral changes could be misinterpreted by the human. In these examples, misunderstanding of intent could cause a number of safety issues relating to operating in close proximity, as well as a degradation in trust [34].

A great deal of work by the RCTA has gone into evaluating different processes for both human-to-robot and robot-to-human communication. One important area of research worked towards AI that can understand human communication in order to minimize the amount of training needed to integrate a robotic system into a team. This includes ways to sense a communication signal as well as how to interpret that sensed signal and translate it into robotic behaviors that can be executed in the physical environment.

Gestures are used by people for communication, such as pointing to locations or waving another driver to perform a driving task, as mentioned earlier. The Army uses standard gestures to convey instructions across distances (e.g., US Army Field Manual 21–60) [35]. The RCTA has researched the use of human-to-robot gestures, how to sense them, and translate in robot behaviors (for a complete review of the literature, see [36]), as well as robot-to-human communication through vestibular and tactile feedback [37].

Speech is another mode of human-to-robot communication that was researched by the RCTA. Specifically, it was essential to quantify the vocabulary used by Soldiers to command a robot to execute tasks. The main goal of this research was to find out how actual users would speak versus how robot programmers and technology developers think about the words to use [38]. Beyond speech, vocabulary is the area of natural language that is of greatest importance. Work is ongoing to use natural language for communication with robots. This is a difficult problem, as it is critical to both understand the words being spoken, but also to infer the intended meaning of the words. Duvallet et al. [39], for example, addressed how to infer maps and behaviors from natural language. Robot-to-human speech-based communication was also investigated by looking at text-based communication via a user display. Text-based information was primarily used to communicate a robot's need for more information or to request support to disambiguate mission needs.

A primary outcome of all this research was the development and assessment of a multimodal interface that incorporated gesture, speech, and a visual display for communication to support human-to-robot commands and robot-to-human requests for assistance [40, 41]. Although these multimodal studies on bidirectional communication do not directly measure trust, the implication is that two-way communication that provides the ability to clarify instructions, ask for additional information, or make suggestions beyond simple constructs, will all add to developing richer interactions and facilitate trust.

4.2   Transparency

Transparency of behaviors, communication, and intent can help calibrate trust in a human-robot team. Transparency is defined as "the descriptive quality of an interface pertaining to its abilities to afford an operator's comprehension about an intelligent agent's intent, performance, future plans, and reasoning process" [42]. It is a widely researched topic for improving communication of intent reasoning for joint decision-making by the wider research community. For the RCTA, the informational needs of mutually interdependent human-agent teams must be anticipated if these teams are to act effectively. Transparency research is needed to identify how information should be presented and communicated between team members to develop shared understanding.

One of the most substantial outcomes of the RCTA work has been the development of the Situation Awareness agent-based Transparency (SAT) model [42]. It builds on the SA models developed by Endsley (1997) but rather than focus on the cognitive requirements for differing levels of SA, the SAT model looks at transparency requirements needed to understand the intelligent agent's task parameters, logic, and predicted outcomes. The three levels tell you what's going on and what the agent is trying to achieve (SAT Level 1), the reasoning process for the agent's decision (Level 2 SAT), and what the operator should expect to happen in the future (Level 3 SAT). Identification of the correct SAT level for mediating transparency can support the process of trust calibration supporting appropriate trust-based outcomes.

Another effort of the RCTA was to understand the role of transparency during the trust process (see Fig. 1). Transparency in terms of robot-to-human communication modality and amount of transmitted contextual information have been suggested to be factors that can affect trust development. Findings suggested that different modalities (text, auditory, pictures) had no overall degradation effects and that contextual understanding was key to trust development [43]. More so, too much or too little information both had negative implications on trust. The amount of robot-to-human information cannot only affect trust, but also is linked to complacency whereby too much information can add ambiguity in the human team member's interpretation of the robot's actions. This ambiguity encourages the trust-based outcome of complacency and results in poorer performance, while direct access to a robot's reasoning process for decision-making has been shown to increase trust and reliance on the robot [44].

5   Social Interactions

With the instantiation of a teaming requirement also comes a change in the role-based structure of the interaction from teleoperation to collaborative decision-making [11]. What this means is that human and robots will be having more social interactions that require the capability to operate in close proximity, participate in collaborative decision-making, and have the capability for back-and-forth or call-and-response type communication. The appropriateness of social interactions are mediated by both cultural and social rules. As a starting point, an individual Soldier and single robot can be thought of as a team, where social cues displayed by the Soldier, such as unconscious head nods, gestures, eye gaze or body postures used during communications and other interactions, might be used to interpret intent, emotion and other human qualities that are seen and interpreted in human-human interaction. Social interactions will also occur amongst more than just a single person and robot, with potentially multiple Soldiers interacting with each other as well as with multiple robots. Therefore, it is necessary to research the social dynamics of groups. In addition to easily observable cues such as uniforms and rank insignia, social cues denoting status and possibly trust that could be observable by robots to include who speaks and when, the spatial location, and relations among individuals within the groups. Cultural understanding, or lack thereof, will be important for military operations in environments with different cultural meanings, customs, and constraints. The social and cultural understandings among entities, both human and technological, impact the development and maintenance of trust.

One major limitation for the development and maintenance of trust in human-robot teams is that robots have difficulty interpreting human social signals from cues observable to them, and that current robot coordination techniques are not intuitive to human partners. It is important for robots to have appropriate social-cognitive mechanisms that allow them to function naturally and intuitively with humans. From their review of the literature, [45] recommended a number of methods for modeling social-cognitive mechanisms in robots centered on modeling perceptual, motor, and cognitive architectures of the robot. These include (a) leveraging ecological approaches; (b) incorporating physical and social affordances; (c) using dynamical modeling and analysis of interaction dynamics; (d) instantiating modal perceptual and motor representations; (e) a coupling of action, perception, and cognition; (f) including motor and perceptual resonance mechanisms; (g) the capability to abstract from modal experiences; and (h) leveraging simulation-based top-down perceptual bias. Further, social cognitive and affective neuroscience can contribute to explaining human behavior to address prior limitations with understanding social cues. These cues can signal the degree of trustworthiness [46].

Equally important as the ability of robots to 'read' human social cues and signals is the capability for the human to perceive and understand social signals that are exhibited by a robot. Social signals tied to proxemics behavior, gaze behavior, and changes in speed were found to directly affect perceptions of intent. For example, within the RCTA, [47] found that proxemics significantly affected perceptions of social presence and emotional state. MacArthur, Stowers, and Hancock [48] found that both proximity and speed can directly impact changes in trust. Moreover, social signals are indicative of socially mindful behavior that can promote positive attributions and feeling of safety [49].

6   Conclusion

The RCTA research has contributed significant advances to the growing understanding of human-robot trust. This work resulted in the identification of a number of gaps and identified specific advances that were made to the fields of mental models, situation awareness, intent communication, transparency, and social-cultural cues. These findings have direct implications for developing effective teaming, AI development, and resolving ambiguity and uncertainty during future human-robot collaborations. One area of future research is assessing behavioral indictors and physiological markers of trust to understand the impact of these above listed topics on trust development real-time. A second area of research is the development of the integration elements between situation awareness needs, with transparency supports, and actual real-world system AI to support adaptive changes in control allocation, decision authority, and communication. A third area of research is the continued development of a multimodal interface for bidirectional communication. All of this current and future research will directly support the development of robotic capabilities that are applicable to a range of mission contexts, robust to changes in the environment, and verifiably worthy of trust by their human teammates.

Acknowledgments. The views and conclusions contained in this document are those of the authors and should not be interpreted as representing the official policies, either expressed or implied, of the Army Research Laboratory or the U.S. Government. The U.S. Government is authorized to reproduce and distribute reprints for Government purposes notwithstanding any copyright notation herein.

References

1. Goodrich, M.A., Schultz, A.C.: Human-robot interaction: a survey. Found. Trends Hum. Comput. Interact. **1**(3), 203–275 (2007)
2. Scholtz, J.: Theory and evaluation of human robot interactions. In: Hawaii International Conference on System Sciences, pp. 10–19. IEEE, Big Island, HI (2003)
3. Adams, B.D., Bruyn, L.E., Houde, S., Angelopoulos, P.: Trust in automated systems literature review (Report No. CR-2003-096). Department of National Defense, Toronto, Ontario, Canada (2003)
4. Billings, D.R., Schaefer, K.E., Llorens, N., Hancock, P.A.: What is trust? defining the construct across domains. In: American Psychological Association Conference, Division 21, Orlando, FL (2012)
5. Schaefer, K.E.: The perception and measurement of human-robot trust. Electronic Theses and Dissertations, 2688. University of Central Florida, FL (2013)
6. Phillips, E., Ososky, S., Jentsch, F.: An investigation of human decision-making in a human-robot team task. Hum. Factors Ergon. Soc. Ann. Meet. **58**(1), 315–319 (2014)
7. Hancock, P.A., Billings, D.R., Schaefer, K.E.: Can you trust your robot? Ergon. Des. **19**(3), 24–29 (2011)
8. Hancock, P.A., Billings, D.R., Schaefer, K.E., Chen, J.Y.C., Parasuraman, R., de Visser, E.: A meta-analysis of factors affecting trust in human-robot interaction. Hum. Factors **53**(5), 517–527 (2011)
9. Schaefer, K.E., Chen, J.Y.C., Szalma, J.L., Hancock, P.A.: A meta-analysis of factors influencing the development of trust in automation: implications for understanding autonomy in future systems. Hum. Factors **58**(3), 377–400 (2016)
10. Norman, D.A.: Some observations on mental models. In: Gentner, D., Stevens, A.L. (eds.) Mental Models, pp. 7–14. Lawrence Earlbaum Associates Inc., Hillsdale, NJ (1983)
11. Phillips, E., Ososky, S., Grove, J., Jentsch, F.: From tools to teammates: toward the development of appropriate mental models for intelligent robots. Hum. Factors Ergon. Soc. Ann. Meet. **55**(1), 1491–1495 (2011)
12. Carroll, J.M., Thomas, J.C.: Metaphor and the cognitive representation of computing systems. IEEE Trans. Syst. Man Cybern. **12**(2), 107–116 (1982)

13. Schaefer, K.E., Sanders, T.L., Yordon, R.E., Billings, D.R., Hancock, P.A.: Classification of robot form: Factors predicting perceived trustworthiness. Hum. Factors Ergon. Soc. Ann. Meet. **56**, 1548–1552 (2012)
14. Schaefer, K.E., Billings, D.R., Hancock, P.A.: Robots vs. machines: identifying user perceptions and classifications. In: Cognitive Methods in Situation Awareness and Decision Support (CogSIMA), pp. 138–141. IEEE, New Orleans, LA (2012)
15. Warta, S.F.: If a Robot did "the robot", would it still be called "the robot" or just dancing? Perceptual and social factors in human-robot interactions. Hum. Factors Ergon. Soc. Ann. Meet. **59**(1), 796–800 (2015)
16. Schaefer, K.E., Adams, J.K., Cook, J.G., Bardwell-Owens, A., Hancock, P.A.: The future of robotic design: trends from the history of media representations. Ergon. Des. **23**(1), 13–19 (2015)
17. Ososky, S., Phillips, E., Schuster, D., Jentsch, F.: A picture is worth a thousand mental models. Hum. Factors Ergon. Soc. Ann. Meet. **57**(1), 1298–1302 (2013)
18. Perelman, B.S., Evans III, A.W., Schaefer, K.E.: Mental model consensus and shifts during navigation system-assisted route planning. Hum. Factors Ergon. Soc. Ann. Meet. **61**(1), 1183–1187 (2017)
19. Cannon-Bowers, J.A., Salas, E., Converse, S.: Shared mental models in expert team decision making. In: Castellan, N.J. (ed.) Current Issues in Individual and Group Decision Making, pp. 221–246. Erlbaum, Hillsdale, NJ (1993)
20. Mathieu, J.E., Heffner, T.S., Goodwin, G.F., Salas, E., Cannon-Bowers, J.: Influence of shared mental models on team process and performance. J. Appl. Psychol. **85**(2), 273–283 (2000)
21. Ososky, S., Schuster, D., Jentsch, F., et al.: The importance of shared mental models and shared situation awareness for transforming robots from tools to teammates. In: XIV SPIE Unmanned Systems Technology (2012)
22. Endsley, M.R.: The application of human factors to the development of expert systems for advanced cockpits. Hum. Factors Ergon. Soc. Ann. Meet. **13**, 1388–1392 (1987)
23. Endsley, M.R.: From here to autonomy: lessons learned from human-automation research. Hum. Factors **59**(1), 5–27 (2017)
24. Endsley, M.R.: Autonomous horizons: system autonomy in the air force – a path to the future (AF/ST TR 15-01), volume 1 Human-Autonomy Teaming. Department of the Air Force, Washington, DC (2017)
25. Schuster, D.A.: The effects of diagnostic aiding on situation awareness under robot unreliability. Electronic Theses and Dissertations. University of Central Florida, FL (2013)
26. Endsley,M.R., Jones, W.M.: Situation awareness information dominance & information warfare (AL/CF-TR-1997-0156). US Air Force Armstrong Laboratory, Wright-Patterson AFB, OH (1997)
27. Schaefer, K.E., Chen, J.Y.C., Wright, J., Aksaray, D., Roy, N.: Challenges with incorporating context into human-robot teaming (TR-SS-17-03). In: AAAI Spring Symposium Series, pp. 347–350. AAAI Publications, Stanford, CA (2017)
28. Schaefer, K.E., Aksaray, D., Wright, J.L., Chen, J.Y.C., Roy, N.: Challenges with addressing the issue of context within AI and human-robot teaming. In: Lawless, W., Mittu, R., Sofge, D. (eds.) Computational Context: The Value, Theory and Application of Context with AI. Springer (In Press)
29. Schaefer, K.E., Perelman, B.S., Brewer, R.W., Wright, J., Roy, N., Aksaray, D.: Quantifying human decision-making: implications for bidirectional communication in human-robot teams. In: Human-Computer Interaction International, Las Vegas, NV (2018)
30. Kadushin, C.: Understanding Social Networks: Theories, Concepts, and Findings. Oxford University Press, New York (2012)

31. Azien, I.: The theory of planned behavior. Organ. Behav. Hum. Decis. Process. **50**, 179–211 (1991)

32. Schaefer, K.E., Brewer, R., Putney, J., Mottern, E., Barghout, J., Straub, E.R.: Relinquishing manual control: Collaboration requires the capability to understand robot intent. In: International Conference on Collaboration Technologies and Systems, pp. 359–366. IEEE, Orlando, FL (2016)

33. Schaefer, K.E., Straub, E.R., Chen, J.Y.C., Putney, J., Evans, A.W.: Communicating intent to develop shared situation awareness and engender trust in human-agent teams. Cogn. Syst. Res.: Special Issue on Situation Awareness in Human-Machine Interactive Systems **46**, 26–39 (2017)

34. Straub, E.R., Schaefer, K.E.: It takes two to tango: Automated vehicles and human beings do the dance of driving – Four social considerations for policy. Transportation Part A: Policy and Practice, Special Issue on Autonomous Vehicle Policy. Elsevier (In Press)

35. U.S. Department of the Army: Visual Signals Field Manual (FM 21-60) (1987)

36. Elliott, L.R., Hill, S.G., Barnes, M.: Gesture-based controls for robots: overview and implications for use by Soldiers (ARL-TR-7715). MDUS Army Research Laboratory, Aberdeen Proving Ground (2016)

37. Mortimer, B.J.P., Elliott, L.R.: Identifying errors in tactile displays and best practice usage guidelines. In: Chen, J.Y.C. (ed.) Advances in Human Factors in Robots and Unmanned Systems, AHFE 2017, Advances in Intelligent Systems and Computing, vol. 595, pp. 226–235. Springer, Cham (2017)

38. Barber, D., Wohleber, R.W., Parchment, A., Jentsch, F., Elliott, L.: Development of a squad level vocabulary for human-robot interaction. In: Shumaker, R., Lackey, S. (eds.) Virtual, Augmented and Mixed Reality Designing and Developing Virtual and Augmented Environments, pp. 139–148. Springer, Cham (2014)

39. Duvallet, F., Walter, M.R., Howard, T., Hemachandra, S., Oh, J., Teller, S., Roy, N., Stentz, A.: Inferring maps and behaviors from natural language instructions. In: Experimental Robotics, pp. 373–388 (2016)

40. Barber, D., Abich IV, J., Phillips, E., Talone, A., Jentsch, F., Hill, S.: Field assessment of multimodal communication for dismounted human-robot teams. Hum. Factors Ergon. Soc. Ann. Meet. **59**(1), 921–925 (2015)

41. Barber, D.J., Howard, T.M., and Walter, M.R.: A multimodal interface for real-time soldier-robot teaming. In: SPIE International Society for Optics and Photonics, Baltimore MD (2016)

42. Chen, J.Y.C., Procci, K., Boyce, M., Wright, J., Garcia, A., Barnes, M.: Situation awareness-based agent transparency (ARL-TR-6905). MDUS Army Research Laboratory, Aberdeen Proving Grounds (2014)

43. Sanders, T.L., Wixon, T., Schafer, K.E., Chen, J.Y.C., Hancock, P.A.: The influence of modality and transparency on trust in human-robot interaction. In: Cognitive Methods in Situation Awareness and Decision Support (CogSIMA), pp. 156–159. IEEE (2014)

44. Wright, J.L., Chen, J.Y.C., Barnes, M., Hancock, P.A.: Agent reasoning transparency: The influence of information level on automation-induced complacency (ARL-TR-8044). MDUS Army Research Laboratory, Aberdeen Proving Ground (2017)

45. Wiltshire, T.J., Barber, D., Fiore, S.M.: Towards modeling social-cognitive mechanisms in robots to facilitate human-robot teaming. Hum. Factors Ergon. Soc. Ann. Meet. **57**(1), 1278–1282 (2013)

46. Wiltshire, T.J., Fiore, S.M.: Social cognitive and affective neuroscience in human–machine systems: A roadmap for improving training, human–robot interaction, and team performance. IEEE Trans. Hum. Mach. Syst. **44**(6), 779–787 (2014)

47. Fiore, S.M., Wiltshire, T.J., Lobato, E.J., Jentsch, F.G., Huang, W.H., Axelrod, B.: Toward understanding social cues and signals in human–robot interaction: effects of robot gaze and proxemic behavior. Front. Psychol. **4**(859), 1–15 (2013)

48. MacArthur, K.R., Stowers, K., Hancock, P.A.: Human-robot interaction: Proximity and speed – Slowly back away from the robot. In: Savage-Knepshield, P., Chen, J.Y.C. (eds.) Advances in Human Factors in Robots and Unmanned Systems. Advances in Intelligent Systems and Computing, vol. 499, pp. 365–374. Springer, Cham (2017)

49. Wiltshire, T.J., Lobato, E.J., Garcia, D.R., Fiore, S.M., Jentsch, F.G., Huang, W.H., Axelrod, B.: Effects of robotic social cues on interpersonal attributions and assessments of robot interaction behaviors. Hum. Factors Ergon. Soc. Ann. Meet. **59**(1), 801–805 (2015)

Human-Inspired Balance Control of a Humanoid on a Rotating Board

Erik Chumacero and James Yang[(⊠)]

Department of Mechanical Engineering,
Texas Tech University, Lubbock, TX 79409, USA
{erik.a.chumacero,james.yang}@ttu.edu

Abstract. We present a stability analysis of the upright stance of a model of a humanoid robot balancing on a rotating board and driven by a human-inspired control strategy. The humanoid-board system is modeled as a triple inverted pendulum actuated by torques at the board's hinge, ankle joint, and hip joint. The ankle and hip torques consider proprioceptive and vestibular angular information and are affected by time delays. The stability regions in different parameter' spaces are bounded by pitchfork and Hopf's bifurcations. It is shown that increasing time delays do not affect the pitchfork but they shrink the Hopf bifurcations. Moreover, the human-inspired control strategy is able to control the upright stance of a humanoid robot in the presence of time delays. However, more theoretical and experimental studies are necessary to validate the present results.

Keywords: Humanoid robot · Stability analysis · Upright stance
Delay differential equation · DDE-BIFTOOL

1 Introduction

Despite the popular belief, the development of humanoid robots (HRs) is not new and can be traced back to da Vinci's robot [1], or even earlier. However, the progression of the modern robot began around the 1930's with a robot named Eric, the first HR able to bow, raise, and give a speech. Nowadays HRs have achieved great progress and they are able to perform far more complex tasks, such as walk, run, and jump, among others. However, humanoids are far from human capabilities and there is a great room for improvement. Among many others, an open problem for HRs is related to the ability to remain in quiet upright stance (UPS) and avoid falls, especially when disturbances are present or when standing on translating or rotating boards. Hence, it is necessary to understand how to control the quiet UPS of the HR to prevent falls [2].

A common source of inspiration in engineering problems is found in nature. The human upright posture is regulated by the central nervous system, which processes the proprioceptive and vestibular angular information in order to stimulate the appropriate muscles to accelerate or decelerate the body's center of mass (CoM) and avoid falls [3]. This process is relatively simple and highly efficient, consequently, there is an interest in mimicking it in order to control the humanoid robot's upright stance (HRUPS).

© Springer International Publishing AG, part of Springer Nature 2019
J. Chen (Ed.): AHFE 2018, AISC 784, pp. 115–126, 2019.
https://doi.org/10.1007/978-3-319-94346-6_11

Several approaches have been proposed for controlling the HRUPS. The most popular approach is the manipulation of the zero-moment point, which combines kinetic and kinematic information to indirectly control the HR's CoM [4–6]. This method is relatively simple in design and application [5] but it cannot be utilized when the HR is standing on a compliant board or when there is not full feet-ground contact [7]. In such cases, linear feedback control (mainly proportional-derivative (PD) control) is used to regulate the deviation of the HR from the UPS. A second approach makes use of optimal control in order to let the projection of the HR's CoM lies within the limits of the base of support [8–10]. Another approach uses model-predictive control, where the driving torque minimizes an objective function over a time horizon [11, 12, 13]. PD control is the most commonly used but there is not agreement about the value of the feedback gains, which are usually chosen heuristically [4]. For instance, the proportional (P) gain must be large enough to compensate for the torque due to gravity but small enough to avoid the feet lifting off the ground [4]. Similarly, the selection of the derivative gain is critical because derivative control is highly sensitive to noise and to sudden changes in the controlled variable. Consequently, it is necessary to find ways to avoid heuristic selection of PD gains that render stable the HRUPS.

On the other hand, most HRUPS controllers do not consider vestibular information in the control loop [14]. Vestibular information is critical for human's upright posture control [15–17]. In fact, people with vestibular loss tend to show larger sway and are more prone to falls [16]. Hence, in order to improve HRUPS control by mimicking human's posture control, vestibular angular information is added to the control loop [4, 14]. This is equivalent to adding a term to the feedback loop that considers the HR's position in a global frame of reference. Therefore, there are two P feedback gains, namely the proprioceptive and the vestibular gains. The question now is how to choose the sensory weights of these two P gains in order to achieve stability of the HRUPS.

Finally, it is known that time delays negatively impact the stability of any dynamical system [18]. For instance, Peterka found that the UPS of a HR controlled by a PD control is unstable for a time delay larger than 36 ms [4]. In fact, a time delay in the control loop has a similar effect as a derivative gain [19, 20]. Despite these negative effects, time delays are usually neglected in the HRUPS's stability analysis.

In this paper, we consider vestibular and proprioceptive feedback in the HRUPS's control loop in order to improve the HR's stance control [4]. Furthermore, we consider time delays in the angular information feedback. Particularly, we are interested in determining the effect of varying time delays, proprioceptive gains, and vestibular gains on the stability of the HRUPS. This is done by representing the HR balancing on a rotating board (which will now be referred to as the HR-board system) as a triple inverted pendulum composed of the board, the legs, and the ensemble head-arms and trunk (HAT). The HR-board system is driven by 3 torques: One at the board's hinge, one at the ankle, and one at the hip joint. The board's torque is produced by a torsional spring. The torques at the ankle and hip joints result from the combination of a passive and an active torque, the former due to the stiffness and viscosity of the joints, and the latter due to the action of actuators attached to the joints and affected by time delays.

The closed-loop system (HR-board/control) results in a delayed differential equation (DDE) that is analyzed in a systematic way by using the software DDE-BIFTOOL® [21, 22]. The software has been used with success in the analysis of human upright posture stability [23–27] but not in HRUPS stability. We focus on finding the stable regions (bounded by pitchfork and Hopf bifurcations) in some selected parameter spaces.

The remainder of this paper is organized as follows. The mathematical model of the HR-board system, the proposed controllers, and the corresponding closed-loop system are described in Sect. 2. The study protocol, results, and discussion are given in Sect. 3. Finally, conclusions and future work are given in Sect. 4.

2 System Modeling: Control and Stability Analysis

2.1 Model of the HR-Board System

A HR is composed of multiple degrees of freedom that provide a mobility similar to that of a human body. However, in order to study HRUPS stability, only the most relevant segments and joints are necessary. Here, the HR is considered as a double inverted pendulum and the rotating board as a single inverted pendulum swinging in the sagittal plane [11, 28, 29]. Hence, the HR-board system is represented by a triple inverted pendulum whose corresponding equation of motion is:

$$\mathbf{D}(\mathbf{q})\ddot{\mathbf{q}} + \mathbf{C}(\mathbf{q}, \dot{\mathbf{q}}) + \mathbf{G}(\mathbf{q}) = \mathbf{M} \tag{1}$$

In Eq. (1), $\mathbf{q} = \begin{bmatrix} q_b & q_a & q_h \end{bmatrix}^T$, $\dot{\mathbf{q}}$, and $\ddot{\mathbf{q}}$ are the joint angular position, velocity and acceleration vectors, respectively. Details of the inertia ($\mathbf{D}$), Coriolis ($\mathbf{C}$), and gravity ($\mathbf{G}$) terms are given in the Appendix. $\mathbf{M} = \begin{bmatrix} M_b & M_a & M_h \end{bmatrix}^T$ is the torque vector (see Fig. 1).

2.2 Control Strategy

Control Design and Closed Loop System

The torques at the ankle and hip joints result from the combination of a passive and an active torque. The passive torques are given by PD controllers in terms of the joints' stiffness (K) and viscosity (B) in Eq. (2).

$$\begin{aligned} M_{a,pass}(t) &= -B\dot{q}_a(t) - Kq_a(t) \\ M_{h,pass}(t) &= -B\dot{q}_h(t) - Kq_h(t) \end{aligned} \tag{2}$$

The active torques are given by delayed P controllers that consider ankle's and hip's proprioceptive and vestibular angular information. The time delays are assumed due to the actuators' dynamics and transmission and processing lapses.

$$\begin{aligned} M_{a,act}(t, \tau_a) &= -K_{pa}q_a^a - K_{va}\left(q_b^a + q_a^a\right) \\ M_{h,act}(t, \tau_h) &= -K_{ph}q_h^h - K_{vh}\left(q_b^h + q_a^h + q_h^h\right) \end{aligned} \tag{3}$$

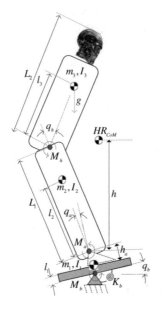

M_b, M_a, M_h : *Torques at the board, the ankle, and the hip.*

q_b, q_a, q_h : *Angles of the board, the ankle and the hip.*

h_a : *Distance from the ankle's joint to the top side of the board.* L_i : *Length of the legs (i=1) and the trunk (i=2).*

l_1 : *Distance from the board's hinge to the CoM of the compound board-feet.*

l_2 : *Distance from the ankle joint to the leg's CoM*

l_3 : *Distance from the hip joint to the HAT's CoM.*

m_i : *Mass of the board-feet (i=1), the legs (i=2) and the HAT (i=3) with corresponding moments of inertia (I_i).*

HR_{CoM} : *CoM of the whole humanoid*

Fig. 1. HR swinging on a rotating board in the sagittal plane.

where we define the notation $q_i^j = q_i(t - \tau_j)$. K_{pa} and K_{ph} are the proprioceptive gains. K_{va} and K_{vh} are the vestibular gains. Finally, τ_a and τ_h are the time delays. Subscripts a and h stand for ankle and hip, respectively. The total torques at the ankle and the hip are described in Eq. (4)

$$\begin{aligned} M_a(t, \tau_a) &= M_{a,pass}(t) + M_{a,act}(t, \tau_a) \\ M_h(t, \tau_h) &= M_{h,pass}(t) + M_{h,act}(t, \tau_h) \end{aligned} \qquad (4)$$

The board's torque hinge is due to a torsional spring of stiffness K_b.

$$M_b(t) = -K_b q_b(t) \qquad (5)$$

Substituting (2)–(5) into (1), defining $\mathbf{x}_1 = \mathbf{q}$, $\mathbf{x}_2 = \dot{\mathbf{q}}$, $\tau = [\tau_a\ \tau_h]^T$, and $\eta = \left[K_b\ C K\ K_{pa}\ K_{ph}\ K_{va}\ K_{vh}\right]^T$ yields the closed-loop system

$$\dot{\mathbf{x}} = \begin{bmatrix} \dot{\mathbf{x}}_1 \\ \dot{\mathbf{x}}_2 \end{bmatrix} = \begin{bmatrix} \mathbf{x}_2 \\ \mathbf{D}^{-1}(\mathbf{x}_1)(\mathbf{M}(t, \tau_a, \tau_h) - \mathbf{B}(\mathbf{x}_1, \mathbf{x}_2) - \mathbf{G}(\mathbf{x}_1)) \end{bmatrix}, \qquad (6)$$

Which can be written in the form of a DDE as

$$\dot{\mathbf{x}} = \mathbf{f}(\mathbf{x}(t), \mathbf{x}(t - \tau), \eta). \qquad (7)$$

Equation (7) has a fixed point at $\mathbf{x}^* = \mathbf{0}$ (quiet UPS) for any positive entry of η and τ.

Control Parameters

For normalizing purposes, the ratios $r_b = K_b/K_b^{cr}$, $r_{pa} = K_{pa}/K^{cr}$, $r_{va} = K_{va}/K^{cr}$, $r_{ph} = K_{ph}/K^{cr}$, and $r_{vh} = K_{vh}/K^{cr}$ are defined: where K^{cr} and K_b^{cr} stand for the critical stiffness at the ankle and hinge respectively and $r_i \geq 0$. K^{cr} is the ankle's critical stiffness that is required to stabilize the rigid-hip HR while standing on a rigid ground. It is defined as $K^{cr} = mgh$, where m is the mass of the HR and h the height of its CoM. In general, each segment of the HR has its own critical stiffness. But in this study, one equivalent critical stiffness between the two HR segments is used. Similarly, K_b^{cr} is the stiffness of the board's torsional spring required to stabilize the HR-board system rotating only about the board's hinge. It is defined as $K_b^{cr} = K^{cr} + m_1 g l_1$, where m_1 is the mass of the board and l_1 the height of its CoM.

Stability Analysis

DDEs usually emerge when time delays are considered in the feedback loop [30]. The DDE given by Eq. (7), linearized at the fixed point $\mathbf{x}^* = \mathbf{0}$ has the characteristic equation described in Eq. (8).

$$\Delta(\lambda) = \lambda\mathbf{I} - \mathbf{A}_a e^{-\lambda\tau_a} - \mathbf{A}_h e^{-\lambda\tau_h} = 0 \qquad (8)$$

where λ is the Laplace's variable and

$$\mathbf{A}_a = \left.\frac{\partial \mathbf{f}(\cdot)}{\partial \mathbf{x}(t-\tau_a)}\right|_{\mathbf{x}=\mathbf{x}^*} \quad \text{and} \quad \mathbf{A}_h = \left.\frac{\partial \mathbf{f}(\cdot)}{\partial \mathbf{x}(t-\tau_h)}\right|_{\mathbf{x}=\mathbf{x}^*} \qquad (9)$$

The characteristic equation $\Delta(\lambda) = 0$ has infinite roots, but only a finite number of them (possibly zero) have a positive real part. The finite number of roots determines whether the UPS is stable. Moreover, the location of the roots in the complex plane depends on the entries of η and τ. Consequently, variations of these entries may produce local bifurcations, which occurs when one or more roots cross the imaginary axis. In this study, we focus on pitchfork (PB) and Hof bifurcations (HB), which can be either supercritical or subcritical. In a supercritical/subcritical PB, a stable/unstable fixed point becomes unstable/stable while other stable/unstable fixed points emerge close by. In a super-critical/subcritical HB, a stable/unstable fixed point becomes unstable/stable while a stable/unstable limit cycle emerges around the fixed point. In mathematical terms, a PB occurs when a single real root passes through the origin of the complex plane. A HB occurs when a pair of complex conjugate roots pass through the imaginary axis. More details about local bifurcations in DDEs are given in [31–33].

Because of Eq. (8) is in matrix form, it is not trivial to determine the roots and numerical methods are needed. DDE-BIFTOOL® in MATLAB®, is a bifurcation analysis software that helps in performing such calculations [22, 34]. It implements continuation of steady-state solutions and computes their stability properties. It also computes steady-state folds, PB and HB. And from HB, it can find branches of periodic solutions, folds of periodic orbits, and torus bifurcations [22, 35].

3   Results and Discussions

Equation (7) is analyzed in a systematic way using DDE-BIFTOOL®. The stability regions for $\mathbf{x}^* = \mathbf{0}$ in 4 sub-sets of $\boldsymbol{\eta}$ and $\boldsymbol{\tau}$ are obtained. Four entries of $\boldsymbol{\eta}$ (K_{pa}, K_{ph}, K_{va}, K_{vh}) and the two entries of $\boldsymbol{\tau}$ (τ_a and τ_h) vary while the other entries remain the same such as $B = 0.3K^{cr}$, $K = 0.75K^{cr}$, and $K_b = 1.5K_b^{cr}$. We tested the combinations detailed in Table 1.

Table 1. Varying and fixed parameters tested in each combination.

Combination #	Varying parameters	Fixed parameters
1	K_{pa}, K_{va}, τ_a	$K_{ph} = K_{vh} = 0.5K^{cr}, \tau_h = 0$
2	K_{pa}, K_{va}, τ_h	$K_{ph} = K_{vh} = 0.5K^{cr}, \tau_a = 0$
3	K_{ph}, K_{vh}, τ_a	$K_{pa} = K_{va} = 0.5K^{cr}, \tau_h = 0$
4	K_{ph}, K_{vh}, τ_h	$K_{pa} = K_{va} = 0.5K^{cr}, \tau_a = 0$

The results of combinations 1 to 4 are shown in Figs. 2, 3, 4 and 5 respectively, where the stable (green) and unstable (other colors) regions are shown. The number of roots with positive real part is depicted in each region (0 for stable, 1, 2, or 3 for unstable). These regions are bounded by the supercritical PB, supercritical HB (SpHB), or subcritical HB (SbHB).

In Fig. 2, for a point in the K_{pa}, K_{va}, and τ_a parameter space located in the stable region, decreasing K_{pa}, while $K_{va} < 0.9K^{cr}$ remains constant, yields loss of stability via a PB. Similar behavior is found when K_{va} is decreased while $K_{pa} < 2K^{cr}$ remains constant.

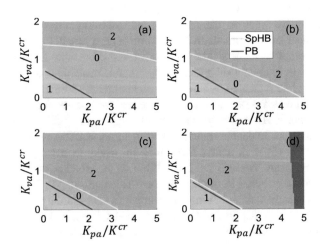

Fig. 2. Stable and unstable regions in the parameter space K_{pa}, K_{va}, and τ_a for time delay τ_a as: (a) 0 ms; (b) 30 ms; (c) 60 ms; (d) 100 ms.

The negative slope of the PB curve suggests that a reduction of K_{pa} can be counteracted by an increment of K_{va}, and vice versa. On the other hand, starting in the stable region, increasing either K_{pa} or K_{va} results in losing stability via a SpHB. This implies the creation of stable limit cycles. In general terms, the stable region is bounded by the SpHB from above and by the supercritical PB from below. On the other hand, increasing τ_a shrinks the area of the stable region until it almost vanishes at about 100 ms (Fig. 2d). The results shown in Figs. 2 and 3 are similar, however, the latter shows that the system preserves the stable region for $\tau_h \geq 180$ ms.

All of the plots shown in Figs. 2 through 5 show that increasing time delays has mainly two effects: (1) shrinking of the region of stability and (2) creation of HBs. Increasing time delays produces SbHB when the ankle's proprioceptive and vestibular gains are considered. However, when the hip's proprioceptive and vestibular gains are taken into consideration, time delays induce the creation of SbHB.

Several highlights can be found from the results. First, increasing either K_{pa} or K_{va} will eventually produce oscillations for the equilibrium stance. These oscillations can be interpreted as postural sway, which is a signal of deficient balance control in humans [36]. However, it was not found if sway also reflects deficient balance in HRs. Figures 2 and 3 suggest that the effects of time delays at the ankle and hip joints are similar only when the ankle's feedback gains (proprioceptive and vestibular) are varied. This supports the practice of considering a global time delay in the stability analysis of HRUPS when only ankle strategy is considered. However, when the hip joint is considered, the dynamics of the system become more complex. In fact, it can be seen from Figs. 4 and 5 that the two time delays have different effects on the HRUPS stability. The ankle's time delay produced the SbHB, which sets a lower limit for K_{ph} with almost any vestibular gain in the stable range (See Fig. 4). This suggests that proprioceptive gain must be high enough to avoid instability of the HRUPS. Contrarily, the hip's time delay produced the SbHB that sets an upper limit for K_{ph}. Considering both results, the values

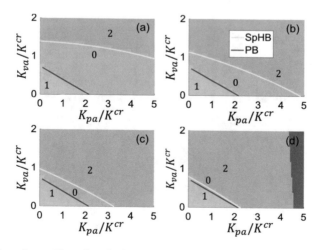

Fig. 3. Stable and unstable regions in the parameter space K_{pa}, K_{va}, and τ_h for time delay τ_h as: (a) 0 ms; (b) 50 ms; (c) 120 ms; (d) 180 ms.

of proprioceptive gains are not only bounded from above, as expected, but also from below. Concerning K_{vh}, Figs. 4 and 5 show that it is limited from below by the PB for any τ_a and τ_h, and from above by the HB for $\tau_h > 120$ ms. This suggests that K_{vh} can be as large as necessary when time delays are short enough. On the other hand, the negative effect of the time delays on the HRUPS stability has been verified by this study. The larger the time delay is, the more difficult it is to find appropriate proprioceptive and vestibular feedback gains to stabilize the HR-board system. This finding is similar to others for different dynamical systems [18, 37,38], where time delays in the control loop shrink the region of stability. Moreover, we have found that the ankle's time delay, compared to the hip's time delay, imposes more restrictive conditions on the selection of K_{pa} and K_{va} (see Fig. 2 and 3).

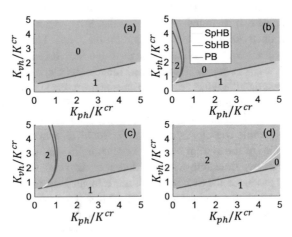

Fig. 4. Stable and unstable regions in the parameter space K_{ph}, K_{vh}, and τ_a for time delay τ_a as: (a) 60 ms; (b) 100 ms; (c) 120 ms; (d) 150 ms.

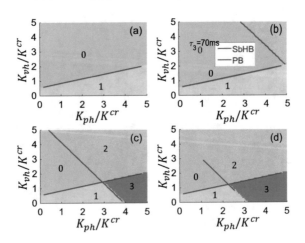

Fig. 5. Stable and unstable regions in the parameter space K_{ph}, K_{vh}, and τ_h for time delay τ_h as: (a) 30 ms; (b) 70 ms; (c) 120 ms; (d) 160 ms.

With respect to the board stiffness, we chose $K_b = 1.5K_b^{cr}$, which produced a stable rigid HR-board system. It would be interesting to reproduce the present study with $K_b < K_b^{cr}$, which, we expect, render an unstable, rigid HR-board system.

4 Conclusions and Future Work

In this study, we developed a mathematical model that describes the dynamics of a HR balancing on a rotating board. The HR is assumed to be actuated at the ankle and hip joints by torques that are proposed as a combination of a passive and an active torque. The passive torque is assumed as given by a PD controller acting without delay due to the stiffness and viscosity of the joints. The active torque is proposed as a delayed P control that considers proprioceptive and vestibular control gains. By means of numerical methods, we predict a wide range of feedback gains and time delays such that the HR-board system is stable. It was shown that the regions of stability in the considered parameter spaces are bounded by PB and HB curves. It was also found that the PB was not affected by the time delays, however, the HB is highly dependent on the time delays. We have proposed a systematic way to select proprioceptive and vestibular control gains, as opposed to an arbitrary selection, such that an unstable dynamic system becomes stable via feedback control.

Despite the interesting results obtained in this study, other questions arise, motivating further study. For instance: What is the subset, in a stable region, for which the produced torques are feasible for a HR? What experiments can be performed in order to validate the behavior described by the proposed model? How valid are the results for a HR given we linearized the system around the fixed point? In conclusion, HR balance control remains an open problem that needs to be addressed from different approaches. Our study is a first step towards a human-inspired HRUPS control and a better understanding of the effects of varying control gains and time delays from a mathematical approach. However, more theoretical and experimental studies should be performed in order to validate these results and to better understand the balance control of HR.

Appendix: Mathematical Model

The following are the detailed expressions for the terms used in Eq. 1 of the main text. The components of the inertia matrix $\mathbf{D}(\mathbf{q})$ are given by Eqs. A1–A7. The terms in the Coriolis vector $\mathbf{C}(\mathbf{q}, \dot{\mathbf{q}})$ are given by Eqs. A8–A10. Finally, the terms in the gravity vector $\mathbf{G}(\mathbf{q})$ are given by Eqs. A11–A13.

$$\mathbf{D}(\mathbf{q}) = \begin{bmatrix} D_{11}(\mathbf{q}) & D_{12}(\mathbf{q}) & D_{13}(\mathbf{q}) \\ D_{21}(\mathbf{q}) & D_{22}(\mathbf{q}) & D_{23}(\mathbf{q}) \\ D_{31}(\mathbf{q}) & D_{32}(\mathbf{q}) & D_{33}(\mathbf{q}) \end{bmatrix} ; \mathbf{C}(\mathbf{q}, \dot{\mathbf{q}}) = \begin{bmatrix} C_1(\mathbf{q}, \dot{\mathbf{q}}) \\ C_2(\mathbf{q}, \dot{\mathbf{q}}) \\ C_3(\mathbf{q}, \dot{\mathbf{q}}) \end{bmatrix} ; \mathbf{G}(\mathbf{q}) = \begin{bmatrix} G_1(\mathbf{q}) \\ G_2(\mathbf{q}) \\ G_3(\mathbf{q}) \end{bmatrix}$$

$$(\text{A1})$$

$$
\begin{aligned}
D_{11}(\mathbf{q}) =& I_1 + I_2 + I_3 + L_1^2 m_1 + L_1^2 m_2 + l_1^2 m_3 - L_1^2 m_2 + L_1^2 m_3 - h_a^2 m_1 + 2l_1 h_a m_2 \\
&+ 2L_2 l_3 m_3 + 2l_1 l_3 m_3 cos(q_a + q_h) + 2l_1 L_1 m_3 cos(q_a) + 2h_a l_3 m_3 cos(q_a + q_h) \\
&+ 2l_1 l_2 m_2 cos(q_a) - L_2^2 m_3 + 2L_1 l_3 m_3 cos(q_h) + 2h_a l_2 m_2 cos(q_a) + 2L_1 l_2 m_2 \\
&+ 2L_1 h_a m_3 cos(q_a) + h_a^2 m_2 + h_a^2 m_3 + 2l_1 h_a m_3
\end{aligned}
\tag{A2}
$$

$$
\begin{aligned}
D_{12}(\mathbf{q}) =& D_{21}(\mathbf{q}) = I_2 + I_3 - L_1^2 m_2 + L_1^2 m_3 - L_2^2 m_3 + 2L_1 l_2 m_2 + L_1 h_a m_3 cos(q_a) \\
&+ 2L_1 l_3 m_3 cos(q_h) + 2L_2 l_3 m_3 + l_1 l_2 m_2 cos(q_a) + h_a l_2 m_2 cos(q_a) \\
&+ l_1 L_1 m_3 cos(q_a) + h_a l_3 m_3 cos(q_a + q_h) + l_1 l_3 m_3 cos(q_a + q_h)
\end{aligned}
\tag{A3}
$$

$$
\begin{aligned}
D_{13}(\mathbf{q}) = D_{31}(\mathbf{q}) =& - m_3 L_2^2 + 2l_3 m_3 L_2 + I_3 + l_1 l_3 m_3 cos(q_a + q_h) \\
&+ h_a l_3 m_3 cos(q_a + q_h) + L_1 l_3 m_3 cos(q_h)
\end{aligned}
\tag{A4}
$$

$$
D_{22}(\mathbf{q}) = I_2 + I_3 - L_1^2 m_2 + L_1^2 m_3 - L_2^2 m_3 + 2L_1 l_2 m_2 2L_2 l_3 m_3 + 2L_1 l_3 m_3 cos(q_h)
\tag{A5}
$$

$$
D_{23}(\mathbf{q}) = D_{32}(\mathbf{q}) = -m_3 L_2^2 + 2l_3 m_3 L_2 + I_3 + L_1 l_3 m_3 cos(q_h)
\tag{A6}
$$

$$
D_{33}(\mathbf{q}) = -m_3 L_2^2 + 2l_3 m_3 L_2 + I_3
\tag{A7}
$$

$$
\begin{aligned}
C_1(\mathbf{q}, \dot{\mathbf{q}}) =& - (l_1 + h_a)(l_2 m_2 sin(q_a) + l_3 m_3 sin(q_a + q_h))\dot{q}_2^2 - (l_1 + h_a)(L_1 m_3 sin(q_a))\dot{q}_a^2 \\
&- 2l_3 m_3(l_1 sin(q_a + q_h) + L_1 sin(q_h))\dot{q}_a \dot{q}_h - 2l_3 m_3(h_a sin(q_a + q_h))\dot{q}_a \dot{q}_h \\
&(-2sin(q_a)(l_1 + h_a)(l_2 m_2 + L_1 m_3) - 2(l_1 + h_a)(l_3 m_3 sin(q_a + q_h)))\dot{q}_b \dot{q}_a \\
&- l_3 m_3(l_1 sin(q_a + q_h) + L_1 sin(q_h))\dot{q}_h^2 - l_3 m_3(h_a sin(q_a + q_h))\dot{q}_h^2 \\
&- 2l_3 m_3(l_1 sin(q_a + q_h) + L_1 sin(q_h))\dot{q}_h \dot{q}_b - 2l_3 m_3(h_a sin(q_a + q_h))\dot{q}_h \dot{q}_b
\end{aligned}
\tag{A8}
$$

$$
\begin{aligned}
C_2(\mathbf{q}, \dot{\mathbf{q}}) =& (l_1 + h_a)(l_2 m_2 sin(q_a) + L_1 m_3 sin(q_a))\dot{q}_b^2 + (l_1 + h_a)l_3 m_3 sin(q_a + q_h)\dot{q}_b^2 \\
&- 2L_1 l_3 m_3 sin(q_h)\dot{q}_b \dot{q}_h - L_1 l_3 m_3 sin(q_h)\dot{q}_h^2 - 2L_1 l_3 m_3 sin(q_h)\dot{q}_a \dot{q}_h
\end{aligned}
\tag{A9}
$$

$$
\begin{aligned}
C_3(\mathbf{q}, \dot{\mathbf{q}}) =& l_3 m_3^2(l_1 sin(q_a + q_h) + L_1 sin(q_h))\dot{q}_1^2 + l_3 m_3^2(h_a sin(q_a + q_h))\dot{q}_b^2 \\
&+ 2L_1 l_3 m_3 sin(q_h)\dot{q}_b \dot{q}_a + L_1 l_3 m_3 sin(q_h)\dot{q}_a^2
\end{aligned}
\tag{A10}
$$

$$
\begin{aligned}
G_1(\mathbf{q}) =& - g(h_a m_3 + l_1 m_1 + l_1 m_2 + l_1 m_3)sin(q_b) - gl_2 m_2 sin(q_b + q_a) \\
&+ L_1 m_3 sin(q_b + q_a) - g(l_3 m_3 sin(q_b + q_a + q_h) + h_a m_2 sin(q_b))
\end{aligned}
\tag{A11}
$$

$$
G_2(\mathbf{q}) = -gsin(q_b + q_a)(L_1 m_3 + l_2 m_2) - gl_3 m_3 sin(q_b + q_a + q_h)
\tag{A12}
$$

$$
G_3(\mathbf{q}) = -gl_3 m_3 sin(q_b + q_a + q_h)
\tag{A13}
$$

References

1. Moran, M.E.: The da Vinci Robot. J. Endourol. **20**, 986–990 (2006)
2. Fujiwara, K., Kanehiro, F., Kajita, S., Kaneko, K., Yokoi, K., Hirukawa, H.: UKEMI: falling motion control to minimize damage to biped humanoid robot. In: International Conference on Intelligent Robots and System. IEEE (2002)
3. Tahboub, K.A.: Biologically-inspired humanoid postural control. J. Physiol. Paris, **103**, 195–210 (2009)
4. Peterka, R.J.: Comparison of human and humanoid robot control of upright stance. J. Physiol. Paris, **103**, 149–158 (2009)
5. Hyon, S., Cheng, G.: Passivity-based full-body force control for humanoids and application to dynamic balancing and locomotion. In: International Conference on Intelligent Robots and System. IEEE (2006)
6. Vukobratovic, M., Borovac, B.: Zero-moment point -thirty five years of its life. Int. J. Humanoid Robot. **1**, 157–173 (2004)
7. Tamegaya, K., Kanamiya, Y., Nagao, M., Sato, D.: Inertia-coupling based balance control of a humanoid robot on unstable ground. In: International Conference on Humanoid Robots. IEEE (2008)
8. Li, Y., Levine, W.S.: An optimal control model for human postural regulation. In: American Control Conference. IEEE (2009)
9. Li, Y., Levine, W.S.: An optimal model predictive control model for human postural regulation.In: Mediterranean Conference on Control and Automation. IEEE (2009)
10. Li, Y., Levine, W.S.: Models for human postural regulation that include realistic delays and partial observations. In: Conference on Decision and Control. IEEE (2009)
11. Aftab, Z., Robert, T., Wieber, P.-B.: Ankle, hip and stepping strategies for humanoid balance recovery with a single model predictive control scheme. In: IEEE International Conference on Humanoid Robots. IEEE (2012)
12. Castano, J.A., Zhou, C., Li, Z., Tsagarakis, N.: Robust model predictive control for humanoids standing balancing. International Conference on Advanced Robotics and Mechatronics. IEEE (2016)
13. Bilgin, N., Ozgoren, M.K.: A balance keeping control for humanoid robots by using model predictive control. In: International Carpathian Control Conference. IEEE (2016)
14. Mergner, T., Schweigart, G., Fennell, L.: Vestibular humanoid postural control. J. Physiol. Paris, **103**, 178–194 (2009)
15. Jeka, J.J., Schöner, G., Dijkstra, T., Ribeiro, P., Lackner, J.R.: Coupling of fingertip somatosensory information to head and body sway. Exp. Brain Res. **113**, 475–483 (1997)
16. Mergner, T.: A neurological view on reactive human stance control. Ann. Rev. Control, **34**, 177–198 (2010)
17. Ishida, A., Imai, S., Fukuoka, Y.: Analysis of the posture control system under fixed and sway-referenced support conditions. IEEE Trans. Biomed. Eng. Inst. Electr. Electron. Eng. **44**, 331–336 (1997)
18. Stepan, G.: Delay effects in the human sensory system during balancing. Philos. Trans. R. Soc. A Math. Phys. Eng. Sci. **367**, 1195–1212 (2009)
19. Atay, F.M.: Balancing the inverted pendulum using position feedback. Appl. Math. Lett. **12**, 51–56 (1999)
20. Peterka, R.J.: Sensorimotor integration in human postural control. J. Neurophysiol. **88**, 1097–1118 (2002)
21. Engelborghs, K., Luzyanina, T., Roose, D.: Numerical bifurcation analysis of delay differential equations. J. Comput. Appl. Math. 265–275 (2002)

22. Sieber, J., Engelborghs, K., Luzyanina, T., Samaey, G., Roose, D.: DDE-BIFTOOL v.3.0 Manual: bifurcation analysis of delay differential equations [Internet] (2016). http://arxiv. org/abs/1406.7144

23. Cruise, D.R., Chagdes, J.R., Liddy, J.J., Rietdyk, S., Haddad, J.M., Zelaznik, H.N., et al.: An active balance board system with real-time control of stiffness and time-delay to assess mechanisms of postural stability. J. Biomech. **60**, 48–56 (2017)

24. Chagdes, J.R., Haddad, J.M., Rietdyk S, Zelaznik HN, Raman A.: Understanding the role of time-delay on maintaining upright stance on rotational balance boards. International Design Engineering Technical Conference. ASME (2015)

25. Chagdes J.R., Rietdyk, S., Jeffrey, M.H., Howard, N.Z., Raman, A.: Dynamic stability of a human standing on a balance board. J. Biomech. **46**, 2593–2602 (2013)

26. Chagdes, J.R, Rietdyk, S., Haddad, J.M., Zelaznik, H.N, Cinelli, M.E., Denomme, L.T., et al.: Limit cycle oscillations in standing human posture. J. Biomech. **49**, 1170–1179 (2016)

27. Cruise, D.R., Chagdes, J.R., Raman, A.: Dynamics of upright posture on an active balance board with tunable time-delay and stiffness. In: International Design Engineering Technical Conference. ASME (2016)

28. Asmar, D.C., Jalgha, B., Fakih, A.: Humanoid fall avoidance using a mixture of strategies. Int. J. Humanoid Robot. **9** (2012). https://doi.org/10.1142/S0219843612500028

29. Jalgha, B., Asmar, D., Elhajj, I.: A hybrid ankle/hip preemptive falling scheme for humanoid robots. International Conference on Robotics and Automation. IEEE (2011)

30. Sieber, J., Krauskopf, B.: Complex balancing motions of an inverted pendulum subject to delayed feedback control. Phys. D Nonlinear Phenom. **197**, 332–345 (2004)

31. Dercole, F., Rinaldi, S.: Dynamical systems and their bifurcations. In: Advanced Methods of Biomedical Signal Processing, pp. 291–325 (2011)

32. Roose, D., Szalai, R.: Continuation and bifurcation analysis of delay differential equations. In: Numerical Continuation Methods for Dynamical Systems. Springer, Dordrecht (2007)

33. Li, X., Ruan, S., Wei, J.: Stability and bifurcation in delay-differential equations with two delays. J. Math. Anal. Appl. **236**, 254–280 (1999)

34. Engelborghs, K.: DDE-BIFTOOL: a Matlab package for bifurcation analysis of delay differential equations (2000)

35. Luzyanina, T., Engelborghs, K., Lust, K., Roose, D.: Computation, continuation and bifurcation analysis of periodic solutions of delay differential equations. Int. J. Bifurc. Chaos Appl. Sci. Eng. **7**, 2547–2560 (1997)

36. Peterka, R.J.: Postural control model interpretation of stabilogram diffusion analysis. Biol. Cybern. **82**, 335–343 (2000)

37. Suzuki, Y., Morimoto, H., Kiyono, K., Morasso, P.G., Nomura, T.: Dynamic determinants of the uncontrolled manifold during human quiet stance. Front. Hum. Neurosci. **10**, 618 (2016)

38. Suzuki, Y., Nomura, T., Morasso, P.: Stability of a double inverted pendulum model during human quiet stance with continuous delay feedback control. International Conference of the Engineering in Medicine and Biology Society. IEEE (2011)

Investigating Ergonomics in the Context of Human-Robot Collaboration as a Sociotechnical System

Daniel Rücker[1(✉)], Rüdiger Hornfeck[1], and Kristin Paetzold[2]

[1] Technische Hochschule Nürnberg, Kesslerplatz 12, 90489 Nürnberg, Germany
{Daniel.Ruecker,Ruediger.Hornfeck}@th-nuernberg.de
[2] Universität der Bundeswehr München,
Werner-Heisenberg-Weg 39, 85577 Neubiberg, Germany
Kristin.Paetzold@unibw.de

Abstract. In this publication, we investigate how the term ergonomics could be defined for human-robot collaboration (HRC) as a sociotechnical system (STS). Thus, we compare different definitions of ergonomics and human factors and conclude on a definition suggested for adoption. Moreover, we compile a list of human factors relevant to that context. We conducted this investigation, because HRC is mainly viewed from a technical viewpoint, although the implications of human involvement should not be underestimated. However, ergonomic evaluation of HRC is based on old methods. The main purpose of this publication is therefore to contribute to a foundation for new ergonomic evaluation methods.

Keywords: Human factors · Human-Robot Collaboration
Sociotechnical Systems

1 Introduction

This publication focuses on regarding all relevant aspects of human factors and ergonomics in the context of Human-Robot Collaboration (HRC) as a sociotechnical system (STS). It is meant as groundwork to establish novel ergonomic evaluation methods for collaborative robots designed for assembly and production systems. This in turn will lead to new approaches in the development of collaborative robots.

HRC as an alternative to full automation and human labor in assembly and production systems seems a promising way to combine the strengths of human and robot, mainly increasing the ergonomics and efficiency of the human activity. However, HRC has been subject to research for over 20 years now, while the market share of HRC robots is still marginal. Besides safety issues, task allocation and ergonomic evaluation are the main challenges.

Until now, research focused on the technical issues, i.e. safety and task allocation while mainly disregarding ergonomic aspects. This makes groundwork like this publication necessary to follow a methodical approach to improve ergonomics in assembly and production systems using HRC.

© Springer International Publishing AG, part of Springer Nature 2019
J. Chen (Ed.): AHFE 2018, AISC 784, pp. 127–135, 2019.
https://doi.org/10.1007/978-3-319-94346-6_12

Regarding ergonomic evaluation, the problems are old methods coming from the evaluation of human labor that are ill adapted to HRC [1]. To enable the development of new methods, a complete and precise definition of ergonomics in the context of HRC must be found. The interpretation of HRC as a STS acts as a boundary for this investigation.

This is meaningful, because the field of ergonomics and human factors has a very broad scope that can be broken down into multiple subtopics. This makes the definition of ergonomics and human factors depending on the regarded context. Therefore, adopting existing definitions without investigating the fitness to the context is not an improvement over the present situation.

While this publication is not presenting a new method or any new data, it lays a very important foundation for future work and should be presented by itself to make the reasoning behind the decision on a definition comprehensible.

After describing the method used to gather relevant data, we present the chosen definitions from related research topics, as well as global and common definitions in the main part of this publication. Following up, we discuss the findings while taking the implications of HRC as a STS into account.

Summing up, this publication tries to answer the following research question:

Regarding Human-Robot Collaboration as a sociotechnical system, how can ergonomics be defined and what aspects of ergonomics have to be considered to enable the ergonomic evaluation of collaborative robots?

2   Method

Building upon the research question, the main task of this publication is twofold: find a suitable definition of the terms human factors and ergonomics and compile a list of aspects that can be used in future publications to define metrics of new methods for ergonomic evaluation.

The investigation of literature is limited by regarding those terms only in the context of HRC as a STS. Therefore, corresponding definitions that have been chosen are presented in the following.

For HRC, we use the definition as a kind of human-robot collaboration from Bütepage and Kragic [2]. They define HRC as an interaction with these properties:

1. Human and robot exchange information
2. Both share a representation of the task
3. Both share the same goal
4. The subtasks performed by both partners are interdependent
5. HRC leads to mutual learning and adaptation, requires mutual trust

As definition of STS, we used the one from Whitworth: "Social-technical systems arise when cognitive and social interaction is mediated by information technology rather than the natural world." [3].

Mariani [4] builds on this definition and describes the following characteristics as essential to STS:

1. Some properties of STS can only be evaluated after deployment, not at design time
2. STS are mostly non-deterministic, i.e. they produce different outputs from the same input
3. At design time it is not possible to predict all ways in which humans may interact with STS
4. Awareness is described as fundamental in STS
5. Adaptation of humans to systems is not exclusive; they may also adapt the systems to them.

To find a definition, we took general definitions of human factors and ergonomics from dictionaries, books on these topics and organizations focusing on human factors and ergonomics. In addition, we searched journals and conferences for entries regarding the definitions they used.

The compilation of a list of ergonomic aspects used the same sources, however the focus is more on current research than on books.

The main search terms were *human, robot, collaboration, sociotechnical, ergonomics, human factors*. The search engines used to find publications were Scopus, Google Scholar, ScienceDirect and Web of Science.

For both tasks, the abovementioned properties of HRC and STS are used to concentrate all findings into one definition and one list of human factors.

3 Results

In the first subsection of this chapter, we state all regarded definitions of ergonomics and human factors, whereas the second lists all sources used to compile a list of human factors important for HRC as a STS.

3.1 Definitions of Ergonomics and Human Factors

The first definitions identified as important are the definitions of both terms from the Cambridge Dictionary. They define ergonomics as:

"the scientific study of people and their working conditions, especially done in order to improve effectiveness" [5] and human factors engineering as:

"the act of studying how people use systems or equipment in order to design, develop, and create technology that is safer, more effective, etc." [6]. It should also be noted that ergonomics is stated under the *see also* section of the human factors engineering definition.

More specific are the definitions of international organizations of ergonomics and human factors.

The International Ergonomics Association (IEA) has the following text under the section *Definition*:

"Ergonomics (or human factors) is the scientific discipline concerned with the understanding of interactions among humans and other elements of a system, and the profession that applies theory, principles, data and methods to design in order to optimize human well-being and overall system performance.

Practitioners of ergonomics and ergonomists contribute to the design and evaluation of tasks, jobs, products, environments and systems in order to make them compatible with the needs, abilities and limitations of people.

Ergonomics helps harmonize things that interact with people in terms of people's needs, abilities and limitations." [7].

The Human Factors and Ergonomics Society [8] uses two definitions on their website "What Is Human Factors/Ergonomics?". While one is the definition of the IEA stated above, the other is from the Computer Ergonomics for Elementary School Students (CergoS):

"Ergonomics and human factors use knowledge of human abilities and limitations to design systems, organizations, jobs, machines, tools, and consumer products for safe, efficient, and comfortable human use." [9].

The definition from the IEA has also been adopted in research publications, with Dul, Bruder et al. being a prominent example. [10].

3.2   Sources for a List of Relevant Human Factors

The first source for human factors is the IEA, which was already quoted for their definition of ergonomics in the section before. They divide ergonomics into three main domains: physical ergonomics, cognitive ergonomics and organizational ergonomics. Although the terms contained in Table 1 are called topics by the IEA, we interpret them as individual or grouped human factors.

Table 1. Human Factors taken from the IEA website [7]. Entries marked with * were excluded for reasons stated in the discussion (Sect. 4).

Physical ergonomics	Cognitive ergonomics	Organizational ergonomics
Working postures	Mental workload	Communication
Materials handling	Decision-making	Crew resource management
Repetitive movements	Skilled performance	Work design
Work related musculoskeletal disorders	* Human-computer interaction	Design of working times
Workplace layout	Human reliability	Teamwork
Safety	Work stress	Participatory design
Health	Training	* Community ergonomics
		Cooperative work
		* New work paradigms
		* Virtual organizations
		* Telework
		* Quality management

Hancock et al. [11] compiled a list of human factors to conduct correlational studies regarding trust in human-robot interaction. They split the factors into three categories, human-related, robot-related and environmental factors.

Table 2. Human factors from Hancock et al. [11]. Entries marked with * were excluded for reasons stated in the discussion (Sect. 4).

Human-related	Robot-related	Environmental
Attentional capacity	Behavior	* In-group membership
Expertise	Dependability	* Culture
Competency	Reliability of robot	Communication
Operator workload	Predictability	* Shared mental models
Prior experiences	Level of automation	Task type
Situation awareness	Failure rates	Task complexity
Demographics	False alarms	Multi-tasking requirement
Personality traits	Transparency	Physical environment
Attitudes towards robots	Proximity	
Comfort with robots	Robot personality	
Self-confidence	Adaptability	
Propensity to trust	Robot type	
	Anthropomorphism	

Ogorodnikova considered human strengths and weaknesses in HRC in her publication [12]. The human factors she mentions can be grouped into three categories: information processing, human error and physical ergonomics. While human error by itself could be regarded as a single human factor, it is listed here as a category with its different forms to represent the significance it has in her publication.

Table 3. Human factors from Ogorodnikova [12].

Information processing	Human error	Physical ergonomics
Decision-making time	Slip	Magnitude of load
Decision accuracy	Lapse	Mutual allocation
Mental workload	Mistake	Dimensions
Vigilance	Violation	Safe methods of operation
Awareness		Kinematics
		Anthropometrics
		Robot anatomy
		Contact forces

Maurice et al. [1] present an approach to reduce musculoskeletal disorders by using collaborative robots. The focus is the ergonomic evaluation of HRC. A main result of the publication is a list of ergonomic indicators used for the evaluation. While these are exclusively biomechanical, we view them as relevant because of their in-depth description.

Table 4. Human factors from Maurice et al. [1]. Entries marked with * were excluded for reasons stated in the discussion (Sect. 4).

Constraint oriented indicators	Goal oriented indicators
Joint normalized position	Balance stability margin
Joint normalized torque	Dynamic balance
Joint velocity	Velocity transmission ratio
Joint acceleration	Force transmission ratio
Joint power	Head dexterity
	Kinetic energy

4 Discussion

In this section, the results of the literature research presented in Sect. 3 are discussed. The section shares the same structure as Sect. 3 for clarity reasons.

4.1 Discussion of Definitions of Ergonomics and Human Factors

While the definitions form the Cambridge dictionary are very general, they are a good way to cross-check definitions from a scientific background against definitions meant for the general linguistic usage. The key points of the definition of ergonomics is the focus on people and their environment and the focus on effectiveness. Compared to the human factors engineering definition, people and effectiveness are used as well, however the focus lies more on the interaction partner than the environment. Even in the Cambridge Dictionary, both terms are marked as related, since ergonomics is remarked below the human factors engineering definition.

The main aspects of the CergoS definition are the focus on human abilities and limitations, the design of different types of systems and safe, efficient and comfortable human use. However, these three focus points are also mentioned by the IEA. Human abilities and limitations is mentioned in the last sentence. The focus on systems is mentioned in the first sentence, although it varies slightly by including the human into the regarded system. The IEA definition wants to optimize human well-being, which can arguably be interpreted to include safe, efficient and comfortable human use of systems.

The IEA definition therefore includes all key points of the CergoS definition and is also widely disseminated and featured in the research domain as well.

Compared to the Cambridge Dictionary definitions, the IEA definition also includes their key points. The focus on people is obvious in both definitions. Whereas Cambridge mentions the environment directly, it can be argued that the IEA includes the human environment into the system as well, if it already includes the human and the other system components.

Therefore, we suggest to adopt the verbal definition of ergonomics and human factors of the IEA in the context of HRC as a STS as well.

4.2   Discussion of Sources for a List of Relevant Human Factors

Since it would go beyond the scope of this publication to investigate every single factor in detail with this number of factors, only the excluded factors are discussed in the context of HRC as a STS, as defined in Sect. 2.

Starting with the IEA list of human factors (see Table 1), all factors from the physical ergonomics category were included.

In the cognitive ergonomics category, the human-computer interaction (HCI) was excluded because it would be misleading, since HCI is a research topic by itself, and its focus points are represented by other human factors, if at all relevant.

Regarding organizational ergonomics, a lot of the stated factors are focused on bigger groups of humans and therefore irrelevant for HRC. If HRC should be regarded at a bigger scale, featuring multiple robots and humans, those factors can be added retroactively. The excluded factors of this category consist of community ergonomics, new work paradigms, virtual organizations, telework and quality management, all sharing the same reasoning for exclusion as stated in this paragraph.

With regard to the factors collected by Hancock et al. (see Table 2), the first two categories, human-related and robot-related, were deemed relevant completed. Naturally, robot and human are the most important aspects in HRC and we deemed none of the factors redundant. In the environmental category, in-group membership, culture and shared mental models were excluded. In-group membership was excluded because it is a very general term for something that is already covered by various other human factors, including communication, teamwork, work design and crew resource management. Culture was excluded for the similar reasons. It is a very general term, and possible implications for HRC are very complex to establish, so in our view, it would be wrong to include it in a baseline human factors list. The same reasoning applies to shared mental models.

Since Ogorodnikova (see Table 3) shares the scope of HRC for her human factors compilation and none of the factors were deemed redundant, no factor was excluded from the three categories. Her work was included because of the relevant factors regarding human error, the factors focusing on information processing missing from other publications and also the mention of contact forces between robot and human.

The ergonomic indicators identified by Maurice, Padois et al. (see Table 4) are intended as parameters for the ergonomic evaluation of HRCs and therefore fit for inclusion into our list of human factors. While they may be very specific by solely focusing on biomechanical measurements of the human, it makes sense to take them into account.

4.3   Compiled List of Relevant Human Factors

In Table 5, the list of all remaining human factors is presented. It has been restructured for better readability. As principal categories, we adopted those of the IEA. Since the physical factors have the most factors by far, they were divided into the subcategories General, Robot-related and Biomechanics. *Operator workload* and *Situation awareness* were removed since they are already represented by *Mental workload* and *Awareness*. *Decision-making* was also removed because *Decision-making time* and *Decision-making accuracy* were already listed.

Table 5. Compiled list of relevant human factors.

Physical ergonomics		
General	Robot-related	Biomechanics
Physical environment	Robot type	Dimensions
Working Postures	Robot anatomy	Kinematics
Materials handling	Robot behavior	Kinetic energy
Repetitive movements	Robot dependability	Magnitude of load
Work related musculoskeletal disorders	Robot reliability	Anthropometrics
Safety	Robot predictability	Contact forces
Health	Robot proximity	Joint normalized position
	Robot personality	Joint normalized torque
	Robot adaptability	Joint velocity
	Level of automation	Joint acceleration
	Failure rates	Joint power
	False alarms	Balance stability margin
	Transparency of action	Dynamic balance
	Anthropomorphism	Velocity transmission ratio
		Force transmission ratio
		Head dexterity
Cognitive ergonomics		
Mental workload	Competency	Propensity to trust
Skilled performance	Prior experiences	Decision-making time
Human reliability	Demographics	Decision accuracy
Work stress	Personality traits	Vigilance
Training	Attitude towards robots	Awareness
Attentional capacity	Comfort with robots	Human error
Expertise	Self-confidence	
Organizational ergonomics		
Workplace layout	Design of working times	Task complexity
Communication	Team work	Multi-tasking requirement
Crew resource management	Participatory design	Mutual allocation
Work design	Task type	Safe methods of operation

5   Conclusion and Future Work

Interpreting HRC as a STS, we chose definitions for both terms as foundation for a literature review on definitions for the terms ergonomics and human factors, as well as different human factors. The goal of the review was to suggest one definition and a list of human factors with relevance to HRC as a STS. Afterwards, we presented definitions and factors gathered during the review. We discussed differences between those definitions, resulting in the suggestion to use the definition of the IEA. Subsequently, the human factors of the different source were discussed, with the reasoning behind the exclusion of factors in the focus. In the last chapter, the compiled list of human factors is presented.

This investigation is intended to contribute to a foundation for the ergonomic evaluation of HRC by suggesting definitions for the essential terms of HRC and a basic list of human factors.

The next step is to find definitions of metrics derived from the human factors list, in order to create a new way to calculate ergonomic scores. Due to the very different natures of the stated factors, we believe this to be a major challenge. While some may represent physical quantities and are therefore calculable, other require empirical methods to develop metrics.

References

1. Maurice, P., Padois, V., Measson, Y., Bidaud, P.: Human-oriented design of collaborative robots. Int. J. Ind. Ergon. **57**, 88–102 (2017)
2. Bütepage, J., Kragic, D.: Human-Robot Collaboration. From Psychology to Social Robotics. CoRR abs/1705.10146 (2017)
3. Whitworth, B.: Social-technical systems. In: Ghaoui, C. (ed.) Encyclopedia of Human Computer Interaction, pp. 533–541. IGI Global, Hershey (2006)
4. Mariani, S.: Coordination of Complex Sociotechnical Systems. Self-organisation of Knowledge in MoK, 1st edn. Springer International Publishing, Cham (2016)
5. Cambridge English Dictionary: ergonomics Meaning in the Cambridge English Dictionary. https://dictionary.cambridge.org/dictionary/english/ergonomics
6. Cambridge English Dictionary: human factors engineering Meaning in the Cambridge English Dictionary. https://dictionary.cambridge.org/dictionary/english/human-factors-engineering?q=human+factors
7. International Ergonomics Association: Definition and Domains of Ergonomics | IEA Website. http://www.iea.cc/whats/index.html
8. Human Factors and Ergonomics Society: What Is Human Factors/Ergonomics? https://www.hfes.org/ContentCMS/ContentPages/?Id=tPCkGvSy4aE=
9. Computer Ergonomics for Elementary School: What is Ergonomics. http://osha.oregon.gov/OSHACergos/ergo.html
10. Dul, J., Bruder, R., Buckle, P., Carayon, P., Falzon, P., Marras, W.S., Wilson, J.R., van der Doelen, B.: A strategy for human factors/ergonomics. Developing the discipline and profession. Ergonomics **55**, 377–395 (2012)
11. Hancock, P.A., Billings, D.R., Schaefer, K.E., Chen, J.Y.C., de Visser, E.J., Parasuraman, R.: A meta-analysis of factors affecting trust in human-robot interaction. Hum. Factors **53**, 517–527 (2011)
12. Ogorodnikova, O.: Human weaknesses and strengths in collaboration with robots. Period. Polytech. Mech. Eng. **52**, 25 (2008)

How Does Presence of a Human Operator Companion Influence People's Interaction with a Robot?

Yuwei Sun[(⊠)] and Daniel W. Carruth

Center for Advanced Vehicular Systems, Starkville, MS, USA
{ys393, dwc2}@msstate.edu

Abstract. It is increasingly common for people to work alongside robots in a variety of situations. When a robot is completing a task, the human operator of the robot may be present at the work site or operating remotely. It is important to understand how people's behavior towards a robot is influenced by the presence or absence of the operator. We observed individuals in public locations as they pass by a robot with and without a visible human operator. We show that individuals were more likely to approach and interact with the robot when it was alone. Also, individuals that interacted with the robot when it was alone were more likely to take extra candy from the robot. Our results suggest that robots with visible operators discourage interaction with the robot but encourage honesty in interactions with the robot.

Keywords: Human factors · Human-robot interaction · Honest behavior

1 Introduction

The market for robotics is thriving with total global sales surpassing $8 billion (USD) in 2015 (IFR 2017). As robots are increasingly common in public areas such as retail or delivery, it is important to better understand the factors that influence interactions between people and robots. Robots are designed in countless styles and for a multitude of functions, both of which can be tailored to the user's purpose. Professional service robots, robots used to complete a commercial task, are the most likely to interact with multiple humans (IFR 2017).

Research on human-robot interaction (HRI) evaluates factors associated with three primary variables: the human, the environment, and the robot. When studying the role of the human in HRI, researchers have focused on background, culture, and demographics (May et al. 2017; Nomura & Sasa 2009; Robinette, et al. 2016). Researchers have studied interactions between humans and robots in different settings to determine how environmental factors affect HRI (Stricker, Muller, & Einhorn 2012). Researchers have also modified robot appearance and behavior to study this aspect of HRI (Kim, et al. 2014).

We are unaware of research examining the effect of the presence or absence of a visible robot operator on humans' responses to the robot. Many robots are directly operated by a human. The operator may be controlling the robot remotely or may be

© Springer International Publishing AG, part of Springer Nature 2019
J. Chen (Ed.): AHFE 2018, AISC 784, pp. 136–142, 2019.
https://doi.org/10.1007/978-3-319-94346-6_13

operating the robot from a position near the robot. In the remote case, individuals interacting with the robot may be unaware that the robot is operated by a human and may believe the robot is autonomous. In the other case, individuals can observe the operator controlling the robot and directly link the robot's behaviors to the individual. In this case, individuals may view the robot as an extension of the operator.

The presence or absence of an operator is likely to alter how an individual views the robot and may affect the individual's likelihood of interacting with the robot and the nature of the interaction. Previous research has shown that when completing a task alone, individuals are more likely to cheat on a task when completing it alone when compared to completing it with another human or a robot in the room (Hoffman, et al. 2016). Participants reported feeling less guilty when monitored by a robot when compared to a human. It is possible that there will be differences in honesty in interactions with a lone robot compared to a robot paired with a human operator. This provides us with an opportunity to examine not only whether the human operator's presence affects individuals' likelihood to interact but also influences the nature of the interaction. In this paper, we describe a research study examining individuals' interactions with a semi-anthropomorphic robot and assessing the effect of the presence or absence of the operator on the observed interactions.

2 Methods

2.1 Participants

We observed a total of 1,450 individuals as they passed through public areas located on the Mississippi State University campus. A total of 991 individuals were observed in the student union and 459 individuals were observed in the library. Most individuals were university students (student demographics: 92% 18–24, 73% white, 21% black, and 6% other).

2.2 Apparatus

We modified a Jaguar V4 tracked robotic platform (DrRobot) with a PVC-pipe frame that approximated a human shape (see Fig. 1). We draped a black fabric skirt around the lower section of the frame. On the arms and torso of the frame, the robot wore a university t-shirt. Above the t-shirt on the 'neck' of the frame, we mounted a Surface Pro 3 tablet. A static image was displayed on the tablet containing text saying "Happy Halloween from CAVS! Please Take One." A platter was fixed to the arms of the frame and a small bowl filled with candy was placed on the platter.

A high-definition camera was mounted under the tablet to observe individual's removing candy from the bowl. The camera was placed so individuals were not able to see the camera. A GoPro camera was mounted above the data collection location and recorded individuals passing through the area.

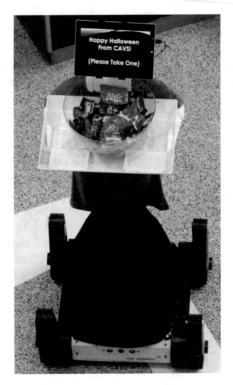

Fig. 1. The assembled Jaguar V4 robotic platform, frame, platter, candy, and tablet.

2.3   Procedure

All procedures were reviewed and approved by the Mississippi State University Institutional Review Board.

Data was collected in two public locations on the Mississippi State University campus: the second-floor lobby of the Mitchell Memorial Library and an open pathway at the Colvard Student Union. Observations occurred on four days: two at the library location and two at the student union location. At each location, three points arranged roughly in an L-shape were designated as P1, P2, and P3. P1 to P2 crossed the direction of pedestrian traffic while P2 to P3 ran parallel to pedestrian traffic. Prior to testing, the candy bowl was filled with 100 pieces of individually wrapped packages of M&Ms (85) and Skittles (15). A small stack of information cards was placed next to the candy. Information about the study was printed on the card. During testing, the robot started at P1, waited 5 min, moved to P2, waited 5 min, moved to P3, waited 5 min, moved to P2, waited 5 min, returned to P1, and waited for five minutes. This pattern was repeated for a total of 60 min or until all the candy was taken (Fig. 2).

In the operator absent condition, the operator was positioned such that the operator could control the robot surreptitiously. In the student union, the operator sat at a table in a large dining area adjacent to the testing location. In the library, the operator controlled

Fig. 2. The library location with position 1, 2, and 3 labeled. The robot began the experiment at position 1.

the robot from an upper floor overlooking the library lobby. To our knowledge, no individual identified the operator during the operator absent condition.

In the operator present condition, the operator positioned himself next to the robot and held the controller in clear view. The operator attempted to limit conversation with individuals when approached by directing them to the information card for details on the robot and the study.

Video of individuals in the general area of the robot and of interactions with the candy were recorded throughout the sessions.

3   Results

The video recordings taken from the location camera and the candy camera were reviewed after completion of all data collection sessions. The location camera was reviewed and the total number of individuals that passed through the data collection area was recorded. Of those that passed through the area, our rater recorded the response of each individual. Individuals that appeared to continue on their initial path with no overt response to the robot were recorded as **ignoring** the robot. Individuals that were observed observing the robot or moving toward the robot but not interacting with the robot or operator were recorded as **curious**. Individuals that interacted with the robot by taking candy and/or an information card were recorded as **interacting**. In the operator present condition, we also recorded whether individuals interacted with only the robot, only the operator, or both.

The candy camera was reviewed and, for each interaction, we recorded whether the individual took one of the packages of candy, as directed by the text on the tablet, or took more than one package of candy.

In the student union location, when the operator was present, a total of 730 individuals were observed in the general area where the robot and operator were located. Of those individuals, 493 (67.5%) appeared to ignore the robot and 206 (28.2%) were observed looking at or appearing to be curious about the robot. Only 23 (3.2%) of the individuals chose to interact with the robot and only 8 (1.1%) chose to interact only with the operator. Of the 23 individuals who interacted with the robot, only 1 (4.3%) was dishonest in their interaction (took more than one candy) and 22 (95.7%) took only one candy.

When the operator was absent and the robot was operating apparently on its own, a total of 261 individuals were observed in the general area where the robot was positioned. Of those individuals, 115 (44.1%) of them appeared to ignore the robot and 78 (29.9%) of them were observed to look at or appeared to be curious about the robot. A total of 68 (26.1%) individuals chose to interact with the robot. Of the 68 people who interacted with the robot, 30 (44.1%) were dishonest in their interaction while 38 (55.9%) acted honestly (Table 1).

Table 1. Observed human behavior at the student union location.

	Operator present	Operator absent
Ignore	67.5%	44.1%
Curious	28.2%	29.9%
Interact	4.3%	26.1%
Honest interaction	95.7%	55.9%

In the library, when the operator was present, a total of 202 individuals were observed in the general area where the robot and operator were located. Of those individuals, 116 (57.4%) chose to ignore the robot and 59 (29.2%) appeared to look at or were apparently curious about the robot. A total of 24 (11.9%) individuals chose to interact with the robot and only 3 (1.5%) individuals interacted with the operator. Of the 24 people who interacted with the robot, only 1 (4.2%) acted dishonestly and 23 (95.8%) acted honestly.

In the library, when the operator was absent and the robot was operating apparently on its own, a total of 257 individuals were observed. Of those individuals, 135 (52.5%) of them chose to ignore the robot and 58 (22.6%) were observed to look at or were curious about the robot. A total of 64 (24.9%) of the individuals chose to interact with the robot. Of the 64 people who chose to interact with the robot, 14 (21.9%) of them acted dishonestly and 50 (78.1%) acted honestly (Table 2).

In the student union, the operator presence had an overall significant effect on individual's behavior, $\chi^2(3, N = 991) = 130.139$, $p < 0.001$. A larger percentage of

Table 2. Observed human behavior at the library location.

	Operator Present	Operator Absent
Ignore	57.4%	52.5%
Curious	29.2%	22.6%
Interact	13.4%	24.9%
Honest interaction	95.8%	78.1%

individuals chose to interact with the robot when the operator was absent than when the operator was present. Operator presence also had an overall significant effect on individual's honesty χ^2 (1, $N = 91$) = 12.102 $p < 0.001$. A larger percentage of individuals acted honestly when the operator was present than when the operator was absent.

In the library, the operator presence also had an overall significant effect on individual's behavior, χ^2 (3, $N = 459$) = 16.272, $p < 0.001$. A larger percentage of individuals chose to interact with the robot when the operator was absent. Operator presence also had an overall significant effect on individual's honesty, χ^2 (1, $N = 88$) = 3.871, $p = 0.49$, with a larger percentage of individuals choosing to act honestly when an operator is present.

4 Conclusions

The result shows that individuals were more likely to approach the robot when it was alone and apparently autonomous than when it was accompanied by an operator. In addition, individuals were more likely to take extra pieces of candy from the robot when it was unaccompanied. There was a difference in dishonesty between the student union and library locations when the operator was absent (55.9% and 78.1%). The union was busier (991 individuals versus 459 individuals) and is a noisier and more active location. The nature of the locations appears to include environmental factors that encourage or discourage dishonest behaviors.

With respect to the difference in interactions and honesty with the operator present, it is likely that when the operator is present and next to the robot, people feel that they are being watched and possibly judged for their interaction with the robot. This may have the effect of both reducing the number of interactions with the robot and encouraging honest interactions.

The results of the study have implications for the use of robots. The presence of a visible operator may discourage some individuals from interacting with the robot. If the goal of the robot is to encourage interaction, as in a retail setting, it will likely be more effective if its operation is autonomous or the operator is not visible.

Acknowledgments. The authors would like to thank the Center for Advanced Vehicular Systems at Mississippi State University for providing funds to support the research project. We would also like to thank Christopher Hudson, T.J. Ciufo, Darren Fray, Austin Chambliss, Jennifer Carruth, and the staff at the Colvard Student Union and the Mitchell Memorial Library for their assistance and support during the study.

References

International Federation of Robotics. Service Robots (2017). https://ifr.org/service-robots/

May, D.C., Holler, K.J., Bethel, C.L., Strawderman, L., Carruth, D.W., Usher, J.: Survey of factors for the prediction of human comfort with a non-anthropomorphic robot in public spaces. Int. J. Soc. Robot. **9**(2), 165–180 (2017)

Nomura, T., Sasa, M.: Investigation of differences on impressions of and behaviors toward real and virtual robots between elder people and university students. In: IEEE International Conference on Rehabilitation Robotics, ICORR 2009, pp. 934–939 (2009). https://doi.org/10.1109/ICORR.2009.5209626

Robinette, P., Li, W., Allen, R., Howard, A.M., Wagner, A.R.: Overtrust of robots in emergency evacuation scenarios. In: ACM/IEEE International Conference on Human-Robot Interaction, pp. 101–108 (2016). https://doi.org/10.1109/HRI.2016.7451740

Stricker, R., Müller, S., Einhorn, E., Schröter, C., Volkhardt, M., Debes, K., Gross, H.-M.: Konrad and Suse, two robots guiding visitors in a university building. In: Autonomous Mobile Systems, pp. 49–58. Springer, Heidelberg (2012)

Kim, R.H., Moon, Y., Choi, J.J., Kwak, S.S.: The effect of robot appearance types on motivating donation. In: ACM/IEEE International Conference on Human-Robot Interaction, pp. 210–211 (2014)

Hoffman, G., Forlizzi, J., Ayal, S., Steinfeld, A., Antanitis, J., Hochman, G., Hochendoner, E., Finkenaur, J.: Robot presence and human honesty: experimental evidence. In: Proceedings of the Tenth Annual ACM/IEEE International Conference on Human-Robot Interaction, pp. 181–188 (2016). https://doi.org/10.1145/2696454.2696487

NAO as a Copresenter in a Robotics Workshop - Participant's Feedback on the Engagement Level Achieved with a Robot in the Classroom

Joseiby Hernandez-Cedeño$^{(\boxtimes)}$, Kryscia Ramírez-Benavides,
Luis Guerrero, and Adrian Vega

University of Costa Rica, San Jose, Costa Rica
{joseiby.hernandez,kryscia.ramirez}@ucr.ac.cr,
luis.guerrero@ecci.ucr.ac.cr, adrvegve@gmail.com

Abstract. Robotics, combined with computer science and human-centered studies, can have a substantial impact in areas such as education and innovation. Robots have proven to be a good tool to gain and maintain users' involvement in different activities. In education, robots can be used as teaching assistants to improve participation, enhance concentration or just to get students' attention. In this research, we involved an NAO, a humanoid robot, in a workshop presentation with the aim of measuring the impact of this technique on the level of engagement showed by the participants. The robot was programmed to simulate speech and gesticulate while it talked to apply the Wizard of Oz technique.

Keywords: Human-robot collaboration · Humanoid robots · Education robots
Human-robot interaction

1 Introduction

We live in a time in which science and technology are a very important part of our daily tasks, and therefore it is too complicated to surprise the youth who have been born in this digital time. Nevertheless, we know that this type of user is filled with ideas, often with innovative ones, that could be important in the development of technology. For this reason, it is important to find different ways to not only motivate them but to maintain their interest in acquiring knowledge that encourages research and development.

This impact we are looking for could be driven by robotics. In this research, robotics was combined with computer science and human-centered studies to achieve this objective. We used an NAO robot during the workshop to try to improve participation, enhance concentration or just get students' attention on a specific subject.

Robotics is an area where several advances and different types of investigations are being made since some time ago. The mechanical metamorphosis that robots have had, has affected the degree of acceptance and its use, making interaction with humans more empathetic [1]. The humans can develop an emotional interaction with robots and have feelings for the machines. Research this area is important to perform the experiment we propose in this publication.

© Springer International Publishing AG, part of Springer Nature 2019
J. Chen (Ed.): AHFE 2018, AISC 784, pp. 143–152, 2019.
https://doi.org/10.1007/978-3-319-94346-6_14

We decided to use robots in the area of education in order to reach and attract the attention of young minds. A humanoid type robot was chosen to interact with a professor during a presentation. This type of robot is usually very effective when dealing with people in tasks that are usually performed by another human. A group of scientists conducted an experiment where a robot interviewed children [2]. The result revealed that the interaction was better than with a human interviewer. The children were more relaxed and they felt more comfortable answering questions. This is the kind of environment that you want to achieve.

The robot during our experiment addresses himself to the public and makes them questions using the Wizard of Oz technique (WoZ). This technique is practically a requirement if you want to be able to develop user-friendly interactive systems. It is important to know why and how to apply the WoZ to obtain the most accurate expected result [3].

We are taking advantage of the capabilities of NAO, to stimulate speech and gesticulate while it talks, for example, because this allows the interactions with humans to seem natural. In addition to this, we are giving the robot social intelligence. Making the robot act like a human, and demonstrating social skills, is important in capturing people's attention and engaging them in a conversation. Recent research shows that social relationships with robots can cause such a positive impact, that their use in education and even therapies for psychological and social aids are strongly recommended [4].

Additionally, an important detail is that we evaluated the level of acceptance of the robot as the professor's assistant using the Godspeed Questionnaire Series (GQS). This has proved to be a very effective method to evaluate robots in different areas. Since there are many experiments that have used it and registered the results, it allows us to compare our research.

There are not many studies that we can find where NAO robots are used to promote learning through the WOZ technique and results are extracted with GQS. But, we can find different studies that support our research of using a robot in the educational context. In 2016, a group of researchers used the NAO to evaluate how they could promote learning in a group of elementary students [5]. Although his approach was based on the social intelligence of the robot, customizing his behavior, the results were positive. This supports and motivates our research.

This paper contains all the details that were related to the experiment. The document is structured as follows: Sect. 2 shows the technical parts of the experiment, the robot and the way it was programmed, the users that participated, and how their perception was measured. Section 3 describes the development of the workshop. Section 4 shows the results obtained with the evaluation of the experiment. In Sect. 5, the results of the evaluation are discussed, and the last Sect. 6 includes the conclusion and the following step in our experiment.

2 Technical Details

This section shows the technical details related to the experiment, describing the entire instruments used in the presentation and analysis of the workshop, as well as a description of the users that participated and the topic that was imparted.

2.1 NAO

The main actor of our experiment is NAO. It is a robot created by SoftBank Robotics [6]. Its characteristics make it a very special and adequate robot for the tests that were made.

It is a humanoid type robot, which means that his figure is like to a human. This robot has arms, head, and legs (Fig. 1 shows NAO in the workshop). It measures 58 cm of height and weighs 5.4 kg. This measure makes it robust and strong, but at the same time, it is friendly to any type of user.

NAO counts with different sensors and devices that allow the interaction in a very natural way. It can walk, maintain the balance and adapt since it counts with legs articulated with 25° of freedom that can move as necessary. In addition, it contains four directional microphones and loudspeakers that helps it communicate, simulating that it talks and listens. The rest of its sensors along with the cameras allow giving NAO the capacity to watch and feel.

One of its most important characteristics that made possible the development of the experiment is that the robot counts with a wireless connection through Wi-Fi, this way it can receive orders without the need of being wired to a device.

Although NAO counts with its own operating system, it recognizes instructions in different programming languages. One of the ways to develop applications for this

Fig. 1. NAO was located in front of the students, right next to the presenter and the presentation. The laboratory was equipped with a computer for each student.

robot is to create a software and import it directly to its internal memory to be executed it in a script way, or by sending the instructions remotely so it can play them. This second option was the one we used to present our workshop.

2.2   Wizard of Oz

The Wizard of Oz technique [7] has been used for many years in different type of experiments. It consists that one user interacts with a certain system that believes it is completely developed; when in fact it is being controlled by a human, in this case, the wizard.

The main objective is to evaluate the interaction of the human with the techno-logical system, in a way that it can be analyzed if the development of the system is feasible or not. Even though in many cases the project might seem like a good idea, the complete implementation could be too expensive and even after testing it to demon-strate that it will not have the expected acceptance. This is the reason why this type of experiments is made, which are generally recorded to be carefully analyzed and obtain the best conclusions.

In our case, the robot was programmed to follow one exact presentation during the workshop. The presentation was prepared with anticipation, and the professor is an expert on the topic being exposed. The order of interactions between the professor and the robot was scripted and practiced in a way that it seemed the most natural possible.

The wizard controlled the robot sending instructions to be executed at the same moment. From a computer connected to the same network as NAO, the software developed allowed the interaction to not be visible to users (Fig. 2 shows an example of the interaction).

The experiment was recorded to be seen and reproduced if necessary.

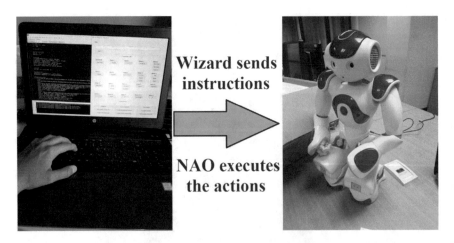

Fig. 2. The interaction between the wizard and NAO is done through a previously developed software program. A graphical interface was created that would allow sending instructions to the robot in an easy way. Through the application, the robot the robot can be instructed to perform tasks such as talking, walking, greeting and even crouching.

2.3   Participants of the Workshop

During the workshop were four different types of users: the students, the professor, and two professor assistants, one was the wizard and the other one was an expert observer. The students were freshman students of computers science of the University of Costa Rica (UCR). The professor organized the workshop and invited 20 students that were concluding their first year to assist.

The professor is part of the school of computer science of UCR and an expert in robotics.

The wizard was presented as the assistant of the workshop, who helped prepared the presentation and helped the professor in anything that was necessary, although he was, in fact, controlling NAO through a desk application that was programmed in advance.

The observer was a professor expert in computing and with knowledge in psychology, which was analyzing the behavior of the students during the presentation.

2.4   Workshop

The workshop was performed in a laboratory of the School of Computing Science of the UCR. The 20 participants were scheduled at a specific hour, each of them was sitting in front of a computer and had a clear vision of where was the NAO as well as the professor's presentation.

The topic presented was an introduction to robotics. General topics and basic concepts were shown, and these were reinforced with NAO.

The robot explained the parts of a robot, such as the sensors and cameras, using his own as an example.

In the final part of the presentation, it was explained how it was possible to develop applications for the robot using different types of programming languages, in a way that the students could think of their own applications in the robotics field.

At the end of the presentation, the professor and NAO asked the students to complete a questionnaire focused on the field of Human-Robot Interaction (HRI), to analyze afterward their perception of the robot.

2.5   Survey

A survey was applied that consisted of two parts: the first was the Godspeed Questionnaire Series (GQS) [8]. This is a tool created with the objective to standardize the way the HRI is measured. The questionnaire helps measure five concepts that are key in HRI: anthropomorphism, animacy, likeability, perceived intelligence and perceived safety.

The answers to each question of the questionnaire are rated on a scale of 5 levels, where 2 are positive, 2 negative and one neutral. In this way, it is easier to perform the evaluation [9].

The questionnaire is available in various languages, and it is free. It might not be the best tool, but it is one of the most used and cited in this field, which allows making comparisons between robots no matter in which field they were tested.

In addition, the questionnaire does not leave aside the factors that may affect the perception of the user towards the robot, like their previous experience with NAO for example. Not only this, but it also measures how clear the participant listened what the robot said and its location according to itself.

The second part of the survey was added as a complement to the GQS, for the experiment it was key to ask if the presence of the robot generated interest in the workshop.

3   Workshop

The best way to prove if the utilization of NAO could attract the attention of the youth or not and increase the level of engagement in a specific topic was to present it in the workshop.

In order for the presentation to be successful, several previous details were analyzed that favored the analysis of the experiment, among these were: the environment, the users, the available material, the topic to be discussed and the speaker.

The workshop was developed in a laboratory in the School of Computer Science of the UCR. The laboratory was ordered in rows, in each row was a maximum of five spaces that could be occupied by the students. The lab was comfortable and with a pleasant weather, this created a comfortable environment for the participants. Besides this, each space had a clear vision of where NAO was located, the presentation and the speaker, which was fundamental to control the attention of the group [10].

The participants of the workshop were students that had just completed the first year of their Computing Science career. These were invited by a UCR professor. The young students were scheduled at a specific hour so that the lab and the materials were ready. The students do not count with experience in the field of programming, but they have a general idea of science and technology.

The available material in the lab was very important since each student counted with a computer that had installed all the necessary programs required to program the applications or send the instructions to NAO.

The workshop topic was focused on robotics. The expert explained the topic in different occasions to other groups in other workshops, but this was the first time NAO was being used as an assistant. The speaker has a Doctorate in Computing Science and is an expert in the field, also he is a professor at UCR.

On the other hand, the presentation showed the historic bases of robotics, as well as its first uses. After this, the main parts of the NAO robot were explained, as well as its characteristics, the details of its sensors, motors, and the rest of the devices the allow it to function and perform different tasks. The wizard sent the instructions to NAO in real time.

After the presentation about robotics was done, the survey was applied. Once completed, the students were asked to give ideas on how an NAO robot can be programmed to interact in an autonomous way as an educational assistant. The ideas given by the students showed how much they were attracted by the presentation and the fact of using a robot as a workshop assistant.

As part of the workshop, it was explained to the students what the real reason for the project and how NAO was programmed, discovering the WoZ. The technique was not detected by the participants beforehand, which caught their attention to the methodology and made them ask questions about this type of investigations.

In order to continue the workshop and motivate even more the students, it was explained to them how to install, program and execute simple actions on the robot. This made the workshop a lot more interesting and appealing. The students used the knowledge acquired during the workshop to present ideas on how to program different applications to use NAO in education and other areas.

The success of the workshop motivated the professors of the School to continue using NAO in the introductory presentations for the new students.

4   Obtained Results

All the students who participated in the workshop completed the survey just after the presentation with NAO. The data obtained were tabulated to be analyzed together with what was perceived by the lecturer and the observer.

The questionnaire, on time, allows us to analyze what was the perception of the students about the NAO in the five key points in which the GQS investigates.

Anthropomorphism. How "human" the robot looks. The results presented that half of the students responded positively and the other half was largely neutral. The students agreed in 60% that NAO seemed "conscious" and "natural", but they were critical and showed that even so the robot still looked machine and artificial.

Animacy. The perception of how alive the robot seems to be. The responses, although they were mostly positive, indicated that the robot was perceived as "lively" and "alive", were not exaggerated. Not one of the students considered the robot to look "organic", which is completely true. Surely, 100% of them considered that NAO was interactive and not at all static.

Likeability. How nice is the robot? Having a positive impression of what NAO looks like is essential so that it catches the attention of the students and they feel more interested in what the robot is presenting to them. This was the feature with the most positive scores. NAO did not receive a single negative rating, 100% of the students considered it "nice", "kind", "pleasant", and "friendly".

Perceived Intelligence. An important point for our experiment. If the goal is to make the students believe that NAO was completely programmed to be an autonomous educational assistant, it is key that they perceive it as an intelligent being. About 40% of the answers were the highest possible rating, obtaining 70% of positive responses. The students rated the robot as a "knowledgeable" and "intelligent" being.

Perceived Safety. This point although it does not directly qualify the robot, instead of that, shows us how users felt when being close to it or interacting with the robot. The type of robot that NAO is has been carefully designed to be accepted by people of all ages. These characteristics are reflected in the results of the questionnaire since in its majority more than 70% of the students answered that they were "relaxed", "calm", but at the same time "surprised".

On the other hand, the survey allows us to obtain data that is directly related to users. All showed to have had some relation with robotics, probably they have read about the subject, investigated, or in their introductory courses in computer science they have seen examples of the type of technology. Only 10% of the students claimed to have had contact with robots, and 90% indicated that they had no prior experience with NAO.

The survey also provides us with data about the students' perception of the environment where they were immersed, the way they listened to the robot, and the distance from it. Regarding the volume, and the timbre of the voice, and the speed of speech, the students were mostly neutral in their responses; there were no negative ratings, but neither positive. The distance of the users and the angle of vision with respect to the robot were slightly negative, this shows us that the students were so captured by the presentation and the robot that they wanted to be closer to NAO.

Although all the parts of the survey allow us to analyze the perception and the level of engagement of the students, the last three questions of this survey let us know exactly the qualification that they give to the workshop.

All the answers regarding the robot related to the presentation were positive. To further specify, 10% of the students showed their neutrality with respect to the presented information, this means that the subject of robotics attracted a great majority. The 40% of the students were strongly convinced that the integration of the robot in the presentation was ideal, plus the other 60% that although they did not give the maximum score at this point, they showed that they liked it. As to whether the presence of the robot generated interest, the students showed that this was essential for the workshop to be a success.

Regarding the opinion of the experts. The exhibiting professor, as he has given the presentation on several occasions, noticed how the NAO integration made his presentation even more interesting towards the students. The level of interest that they had at the end of the workshop was very great and motivating even for the teacher.

The observer noted how the environment was comfortable for the students, and that there was a high level of attention that was provided throughout the presentation. The students were surprised with the first NAO interaction, so they were waiting, waiting when the robot explained details and noticing each of the gestures he made while speaking.

5   Discussion

The results were even more positive than expected. By reviewing the data obtained in the survey, we can extract important details from the students. Although these students are beginning their career in technology they never left aside their scientific and analytical thinking as they qualified the NAO as a robot, a machine, and showed how intelligent they considered it. Moreover, as the most crucial point, it made them analyze the situation in which they were. They did not lose concentration, but on the contrary, they became more interested in the subject that was being exposed.

This interest made the students see the workshop as a gateway to new knowledge. They began to present ideas to be able to use the robot in different projects. At the same time, they understood the challenge that this entails, so they also felt motivated to learn programming languages and to research about HRI.

Seeing the students so committed, the teachers even wished they could repeat the workshop with other students, and give everyone that motivation that could be generated and transmitted.

On the other hand, the observations and the results obtained through the GQS even create more research opportunities. There is a possibility to make comparisons since there are studies that show how NAO has been used in other environments, compared to another type of user, and even being used to perform different tasks [11].

6   Conclusion

Our main conclusion is that the engagement level achieved with a Robot, in this case, NAO, as a copresenter in the classroom, is positively high. Having a robot in class generated such emotion, that it captured all the students from the moment the presentation began.

The results are positive from every point of view. The students rated the robot very well, the presentation and how they felt during the workshop. A key point that made the students feel committed was that they were comfortable with the presence of the robot. NAO was cataloged with a high level of sympathy. This makes us analyze as researchers, since it may be that the success of the project was based on the use of a "cute" robot.

There is a lot of scopes to think about future work. So, from now on, it is planned to repeat the experiment with different types of users, different robots, and even other subjects, they would give us a lot of data to investigate and create new ways to attract the attention of bright minds as young people are.

We know that experiments like these give rise to more research in a field as wide and interesting as HRI. The goal of our research is to benefit everyone who participates in them.

Acknowledgments. This work was partially supported by *Centro de Investigaciones en Tecnologías de la Información y Comunicación* (CITIC), *Escuela de Ciencias de la Computación e Informática* (ECCI) both at *Universidad de Costa Rica* (UCR). Grand No. 834-B7-267. We would like to thank *Programa de Posgrado en Computación e Informatica* and *Sistema de Estudios de Posgrado* at UCR for their support. Additionally, thanks to the *User Interaction Group* (USING) for providing ideas to refine and complete the research.

References

1. Geun, C., Park, J.: From Mechanical Metamorphosis to Empathic Interaction: A Historical Overview of Robotic Creatures. Seoul National University, South Korea (2014)
2. Wood, L., Dautenhahn, K., Rainer, A., Robins, B., Lehmann, H.: Robot-mediated interviews - how effective is a humanoid robot as a tool for interviewing young children? PLoS ONE **8**, e59448 (2013)

3. Dahlbäck, N., Jönsson, A., Ahrenberg, L.: Wizard of Oz Studies—Why And How. Natural Language Processing Laboratory, Sweden (1993)
4. Dautenhahn, K.: Socially intelligent robots: dimensions of human–robot interaction. School of Computer Science, University of Hertfordshire, United Kingdom (2007)
5. Baxter, P., Ashurst, E., Read, R., Kennedy, J., Belpaeme, T.: Robot education peers in a situated primary school study: personalisation promotes child learning. PLoS ONE **12**, e0178126 (2017)
6. Softbank Robotics. https://www.ald.softbankrobotics.com
7. Mäkelä, K., Salonen, E., Turunen, M., Hakulinen, J., Raisamo, R.: Conducting a Wizard of Oz Experiment on a Ubiquitous Computing System Doorman. University of Tampere, Finland (2001)
8. Bartneck, C., Croft, E., Kulic, D.: Measurement instruments for the anthropomorphism, animacy, likeability, perceived intelligence, and perceived safety of robots. Int. J. Soc. Robot. **1**, 71–81 (2009)
9. The Godspeed Questionnaire Series. http://www.bartneck.de/2008/03/11/the-godspeed-questionnaire-series/
10. Shiomi, M., Kanda, T., Koizumi, S., Ishiguro, H., Hagita, N.: Group Attention Control for Communication Robots with Wizard of OZ Approach, Japan (2006)
11. Weiss, A., Bartneck, C.: Meta analysis of the usage of the godspeed questionnaire series. In: Proceedings of the IEEE International Symposium on Robot and Human Interactive Communication, Kobe, Japan, pp. 381–388 (2015)

Motion Planning Based on Artificial Potential Field for Unmanned Tractor in Farmland

Kang Hou[1(✉)], Yucheng Zhang[1], Jinglin Shi[1], and Yili Zheng[2]

[1] Institute of Computing Technology, Chinese Academy of Sciences,
No. 6 Kexueyuan South Road, Zhongguancun, Haidian District,
100190 Beijing, China
{houkang, zhangyucheng, sjl}@ict.ac.cn
[2] Beijing Forestry University, No. 35 Qinghua East Road, Haidian District,
100083 Beijing, China
zhengyili@bjfu.edu.cn

Abstract. A motion planner based on artificial potential field for unmanned tractor is proposed in this paper. In order to get the effective environment model, the model for condition of terrain and impact of ground to tractor is analyzed and built. The tractor usually brings a trailer behind, so the kinematics of tractor with a trailer is presented. The motion planner based on artificial potential field is designed for the unmanned tractor working in farmland. According to the characteristics of the unmanned tractor, the control algorithm and motion planner is optimized. The simulation of the improved motion planner for unmanned tractor is presented and analyzed. And the simulation results are presented to show the effectiveness of the proposed method.

Keywords: Unmanned tractor · Motion planning
Artificial potential field · Tractor-trailer system

1 Introduction

Unmanned tractor is a typical outdoor intelligent unmanned vehicle working in farmland. It helps the farmers do repetitive tasks [1], such as seeding, planting, spraying and weeding, and it can reduce the error factor and the cost of redundancy, improve the efficiency, and accelerate the working procedure. With the rapid development of unmanned vehicles and Internet of vehicles, the relevant technologies [2] also make a great progress, such as vehicular communication network [3, 4], intelligent control and path planning. Tractor, as a kind of ground vehicle, is an important tool for farmers working in the farmland. The scholars all over the world have researched unmanned driving system for the tractor. Matveev [5] proposed two guidance laws to realize automatic path tracking for agricultural tractor. The sliding-mode controller and the combination controller of sliding mode and nonlinear control is adopted to ensure tracking a curve path. Kim [6] developed control actuators to perform the path tracking under the navigation system. Due to the working environment and task, motion planning, a key research field of the unmanned vehicles, is also a challenging problem for unmanned tractor.

J. Chen (Ed.): AHFE 2018, AISC 784, pp. 153–162, 2019.
https://doi.org/10.1007/978-3-319-94346-6_15

Many control algorithms have been adopted by scholars for vehicle unmanned driving system, such as sliding mode, backstopping, neural network and so on. Due to its simplicity and practicability, the artificial potential field (APF) method is widely used for the path planning. An increasing number of researchers have researched the algorithm to plan the robot path. Min [7] proposed a new concept using the APF method to avoid the local minimums occurred and realize the real-time path planning. Cao [8] designed a new APF method based on the relative threat coefficient for robot path planning in an environment with moving target and obstacles. Fan [9] added an exponential factor into APF method to eliminate the chattering phenomenon in the conic curve function model. The working environment of tractor is more complex than the ordinary robot, so the APF method should be optimized to satisfy the tractor motion planning.

In this paper, the motion planning based on artificial potential field for unmanned tractor is presented. The main research contents are as follows. Firstly, the working environments of unmanned tractor in the farmland is researched. The model for condition of terrain and impact of ground to tractor is built and analyzed. Secondly, the kinematics of tractor with a trailer is presented, the model of the tractor and trailer's motion in farmland is built, and the state equation is derived. Thirdly, the artificial potential field algorithm is presented. The motion planner based on artificial potential field is designed for the unmanned tractor working in farmland. According to the characteristics of the unmanned tractor, the control algorithm and motion planner is optimized. Fourthly, the simulation of the improved motion planner for unmanned tractor is presented. In order to get the advantage of the proposed motion planner, the simulation comparison between the proposed method and the traditional control method is performed.

2   Model of Tractor Working Environments

Due to the terrain circumstance and surrounding condition, the working environment of unmanned tractor is different from other unmanned vehicles'. And it brings difficulties for tractor unmanned driving in farmland. For the vehicles, the tires are the only component connecting the vehicle and ground. Especially for the tractor always working in the soft earth, the interaction between the tire and soil is one of the important factors that affect the motion performance of the tractor. According to the characteristics of interaction between the tire and soil, the interaction is analyzed. As shown in Fig. 1, the model of the contact between the tire and soil is built simply [10].

In Fig. 1, the diameter of tire is described as D, and the force acting on the tire is F_Z. The height z_0 between the high contact point and low contact point can be got from

$$z_0 = \sqrt[2n+1]{\left[\frac{F_z\left(\frac{n}{2}+1\right)}{B\sqrt{D}k}\right]^2}. \tag{1}$$

where n and B are tire's coefficients.

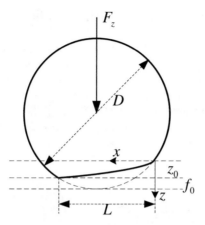

Fig. 1. The model for the contact between the tire and soil.

The deformation component of the tire along the z axis can be got from

$$z(x) = \frac{z_0}{\sqrt{L}} x^{\frac{1}{2}}. \tag{2}$$

The inflation pressure of the tire can be got from

$$p(x) = k \left(\frac{z_0}{\sqrt{L}} \right)^n x^{\frac{n}{2}}. \tag{3}$$

where k is tire's coefficient.

The length L of the contact area can be got from.

$$L = \sqrt{D \cdot z_0}. \tag{4}$$

From the above equations, the contact area and deformation of the tire can be calculated, and they will affect the tractor's motion state.

For the uneven terrain, RMS height can be used to describe the variance of surface roughness, and it is a basic index for measuring surface roughness. The expression of the RMS height is [11]

$$h_{RMS} = \sqrt{\frac{1}{n} \sum_{x=0}^{n} \left[H(x) - \overline{H}(x) \right]^2}. \tag{5}$$

where n is the number of the measurement points, and $H(x)$ is the height of xth measurement point.

The fact is that the terrain condition of the farmland is constantly changing. The roughness of ground surface can be expressed by random array within certain value ranges.

3 Motion Analysis of Tractor with a Trailer

Normally, the tractor has a trailer behind for the farm work, and the trailer can realize some functions, such as plough, sowing, harvest and carry. The trailer tracks the path of the tractor [12]. As shown in Fig. 2, the tractor brings a trailer behind it, and a hinge joint connects them. The front wheels of the tractor provide the motive force and have the steering system, and the rear wheels and the trailer's wheels follows the front wheels. So, the tractor system has two input values for motion control: the linear velocity and direction of the front wheels.

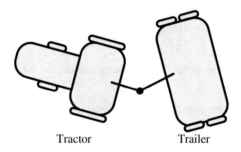

Tractor   Trailer

Fig. 2. A tractor with one trailer.

The coordinate system of the tractor with a trailer is shown in Fig. 3. In the coordinate system, the tractor system has three parts: the front wheels part, the rear wheels part and the trailer's wheels part.

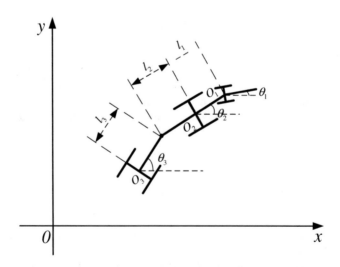

Fig. 3. The coordinate system of the tractor with one trailer.

The states of every part are described as $(x_n, y_n, \theta_n)(n = 1, 2, 3)$ [13]. The coordinate of the wheels' center is $O_n(x_n, y_n)$, and θ_n is the angle between every part direction and the x axis. So, the coordinates of the right and left wheels are respectively $R_n(x_{Rn}, y_{Rn})$ and $L_n(x_{Ln}, y_{Ln})$, where $n = 1, 2, 3$.

The coordinate of rear wheels part is

$$\begin{bmatrix} x_2 \\ y_2 \end{bmatrix} = \begin{bmatrix} x_1 \\ y_1 \end{bmatrix} - \begin{bmatrix} \cos \theta_2 \\ \sin \theta_2 \end{bmatrix} l_1. \tag{6}$$

The coordinate of trailer's wheels part is

$$\begin{bmatrix} x_3 \\ y_3 \end{bmatrix} = \begin{bmatrix} x_2 \\ y_2 \end{bmatrix} - \begin{bmatrix} \cos \theta_2 & \cos \theta_3 \\ \sin \theta_2 & \sin \theta_3 \end{bmatrix} \begin{bmatrix} l_2 \\ l_3 \end{bmatrix}. \tag{7}$$

The coordinate of every part's left wheel is

$$\begin{bmatrix} x_{Ln} \\ y_{Ln} \end{bmatrix} = \begin{bmatrix} x_n \\ y_n \end{bmatrix} + \frac{d}{2} \begin{bmatrix} -\sin \theta_n \\ \cos \theta_n \end{bmatrix}. \tag{8}$$

The coordinate of every part's right wheel is

$$\begin{bmatrix} x_{Rn} \\ y_{Rn} \end{bmatrix} = \begin{bmatrix} x_n \\ y_n \end{bmatrix} + \frac{d}{2} \begin{bmatrix} \sin \theta_n \\ -\cos \theta_n \end{bmatrix}. \tag{9}$$

The two input values of the tractor system are the linear velocity and direction of the front wheels. The tractor's linear velocity is set as v, and the direction of the front wheels is θ_1. So, the derivation of every part's coordinate is

$$\begin{bmatrix} \dot{x}_n \\ \dot{y}_n \end{bmatrix} = \begin{bmatrix} \cos \theta_n \\ \sin \theta_n \end{bmatrix} v \cos(\theta_n - \theta_{n-1}). \tag{10}$$

where $n = 1, 2, 3$, and $\theta_0 = 0$.

The tractor's angular velocity is set as u, so the derivation of front wheels part's angle is

$$\dot{\theta}_1 = u. \tag{11}$$

The derivation of rear wheels part's angle is

$$\dot{\theta}_2 = \frac{v}{l_1} \sin(\theta_2 - \theta_1). \tag{12}$$

The derivation of trailer part's angle is

$$\dot{\theta}_3 = \frac{v}{l_3}\left[\cos\theta_1\sin(\theta_3 - \theta_2) - \frac{l_2}{l_1}\sin\theta_1\cos(\theta_3 - \theta_2)\right]. \tag{13}$$

The state of the tractor with a trailer can be derived by above kinematics formulas, so the model of the tractor and trailer's motion in farmland is built.

4   Artificial Potential Field for Unmanned Tractor's Control

The artificial potential field method is a virtual method, and the motion of a mobile vehicle in the environment can be considered as a movement in an abstract artificial force field. The destination point generates force on the mobile vehicle, and the obstacle generates a repulsive force on the mobile vehicle. Finally, the movement of the robot or vehicle is determined according to the resultant force.

The artificial potential field algorithm is presented. The traditional artificial potential field function is expressed as

$$V(e) = V_a(e) + V_o(e). \tag{14}$$

where $V_a(e)$ represents the attractive potential, and $V_o(e)$ represents the repulsive potential.

So, the virtual force F is expressed as

$$F = F_a + F_o = \nabla V(e) = \begin{bmatrix} \frac{\partial V}{\partial x} \\ \frac{\partial V}{\partial y} \end{bmatrix}. \tag{15}$$

where F_a represents the attractive virtual force, and F_o represents the repulsive virtual force.

$$V_a(e) = ke^2 = k(x - x_d)^2. \tag{16}$$

where k is the attractive gain coefficient, and x_d is the designed value.

$$V_o(e) = j\left(\frac{1}{d} - \frac{1}{d_e}\right)^2. \tag{17}$$

where j is the repulsive gain coefficient, and d_e is the affect distance of obstacle.

So, the derivation of $V_a(e)$ is

$$F_a(e) = 2k(x - x_d). \tag{18}$$

And the derivation of $V_o(e)$ is

$$F_o(e) = 2j\left(\frac{1}{d} - \frac{1}{d_e}\right)\frac{1}{d^2}\nabla d. \tag{19}$$

In order to realize the motion planning for tractor, the motion planner based on artificial potential field is designed for the unmanned tractor [14]. The artificial potential field acting on the unmanned tractor is shown in Fig. 4.

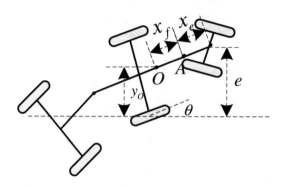

Fig. 4. Artificial potential field based on unmanned tractor.

The optimized artificial potential field function is expressed as

$$V_a(e) = ke^2 = k\left[y_O + \left(x_f + x_e\right)\sin\theta\right]^2 \tag{20}$$

where x_f is the distance between the potential force acting on a vehicle and the center of mass, x_e is the distance between the point of action and the potential force.

So, the virtual force from the potential field is

$$F_a = -\frac{\partial V_a}{\partial e} = -2ke = 2k\left[y_O + \left(x_f + x_e\right)\theta\right] \tag{21}$$

So, the derivation of linear velocity is

$$\dot{v} = \frac{2k\left[y_O + \left(x_f + x_e\right)\theta\right]}{m} \tag{22}$$

And the derivation of angular velocity is

$$\dot{u} = \frac{2k\left[y_O + \left(x_f + x_e\right)\theta\right]}{m}x_f \tag{23}$$

So, the input values of unmanned tractor v and u can be derived for the controller.

5   Simulations and Analysis

In order to verify the feasibility and effectiveness of the proposed method, the above vehicle model and control algorithm are modeled and simulated in Matlab/Simmulink.

In order to get the advantage of the proposed motion planner, the simulation comparison between the proposed method and the traditional control method is performed. PID control is chosen as the traditional control method. The simulations of straight path tracking are carried out respectively using PID and APT method. The tractor's velocity is set as 3 m/s, and the target line is $y = 0$, the coordinate of start position is set as $(0, -2)$. The simulation results of the straight path are shown in Fig. 5.

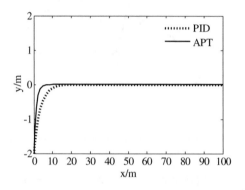

Fig. 5. The simulations of straight path tracking

From Fig. 5 we can get that the APT controller can reach the target line faster than PID controller. Therefore, the APT method has the advantage of short responding time.

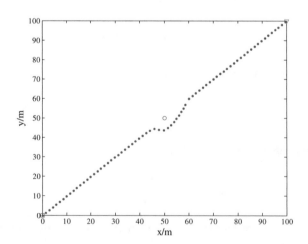

Fig. 6. The simulation of tractor motion planning

In the simulation of tractor motion planning, the work environment is set as a two-dimensional space. The original point [0, 0] is set as the initial position, point [100,100] is set as the target position, and point [50, 50] is set as the position of one obstacle. The simulation result of tractor motion planning using the optimized artificial potential field is shown in Fig. 6.

As shown in Fig. 6, we set that the tractor moves around the obstacle and then fast move to the target point along the original line. This indicates that the proposed artificial potential field can realize the effective and efficient motion for the motion planning of unmanned tractor.

6 Conclusion

In this paper, motion planning for unmanned tractor is presented. The main contributions of this paper include (1) the model of the contact between the tire and soil for unmanned tractor is built, and the motion of tractor with a trailer is analyzed; (2) the motion planner is designed based on artificial potential field algorithm, and the motion planner is improved by considering the characteristics of unmanned tractor. The designed motion planner is not limited to the unmanned tractor working in the farmland. It has the potential for the unmanned vehicle working in the forest.

In future work, we will focus on automation control and group control for unmanned tractors, and the multi-sensor fusion and robust control will be the key topics.

Acknowledgments. This work is supported by National Natural Science Foundation of China (No. 31670719). The authors would like to thank all the members including trainees in ICT/CAS-ASTRI Advanced Wireless Technology Joint Research Center, Institute of Computing Technology, Chinese Academy of Sciences.

References

1. Ishak, W.I.W., Yongwin, L., Razali, M.H.: Development of autonomous bio-production vehicle for agriculture. Int. J. Agric. Sci. **2**(2), 21–27 (2010)
2. Zheng, K., Zheng, Q., Chatzimisios, P., Xiang, W., Zhou, Y.: Heterogeneous vehicular networking: a survey on architecture, challenges, and solutions. IEEE Commun. Surv. Tutor. **17**(4), 2377–2396 (2015)
3. Liu, H., Zhou, Y., Tian, L., Shi, J.: How can vehicular communication reduce rear-end collision probability on highway. In: 2014 IEEE Global Communications Conference, IEEE Press (2015)
4. Zhang, P., Zhou, Y., Liu, H., Tian, L., Shi, J.: A scheme based on graph coloring theory for time slot allocation in integrated VANET-cellular heterogeneous networks. Chin. High Technol. Lett. **26**(6), 550–557 (2016)
5. Matveev, A.S., Hoy, M., Katupitiya, J., Savkin, A.V.: Nonlinear sliding mode control of an unmanned agricultural tractor in the presence of sliding and control saturation. Robot. Auton. Syst. **61**(9), 973–987 (2013)

6. Kim, S.C., Park, W.P., Jung, I.J., Chung, S.O., Lee, W.Y.: Development of unmanned tractor with autonomous and remote control utility. In: 2000 ASAE Annual International Meeting, Technical Papers: Engineering Solutions for a New Century, vol. 1, pp. 2323–2335 (2000)
7. Min, C.L., Min, G.P.: Artificial potential field based path planning for mobile robots using a virtual obstacle concept. IEEE/ASME Int. Conf. Adv. Intell. Mechatron. 2, 735–740 (2003)
8. Cao, Q., Huang, Y., Zhou, J.: An evolutionary artificial potential field algorithm for dynamic path planning of mobile robot. IEEE/RSJ Int. Conf. Intell. Robots Syst. 3331–3336 (2007)
9. Fan, X.P., Li, S.Y., Chen, T.F.: Dynamic obstacle-avoiding path plan for robots based on a new artificial potential field function. Control Theory Appl. 22(5), 703–707 (2005)
10. Schmid, I.C.: Interaction of vehicle and terrain results from 10 years research at IKK. J. Terramech. 32(1), 3–26 (1995)
11. Lu, Z.X., Wu, X.P., Perdok, U.D., Hoogmoed, W.B.: Analysis of tillage soil surface roughness. Trans. Chin. Soc. Agric. Mach. 35(1), 112–116 (2004)
12. Yuan, J.: Hierarchical motion planning for multi-steering tractor-trailer mobile robots with on-axle hitching. IEEE/ASME Trans. Mechatron. 22(4), 1652–1662 (2017)
13. Han, Q., Huang, Y., Yuan, J., Kang, Y.: A method of path planning for tractor-trailer mobile robot based on the concept of global-width. In: Proceedings of the 5th World Congress on Intelligent Control and Automation, vol. 6, pp. 4773–4777. IEEE Press, Hangzhou (2004)
14. Wang, M.L., Chen, W.W., Wang, J.E.: An adaptive automatic correction control method of lane departure based on road artificial potential field. China Mech. Eng. 24(24), 3402–3407 (2013)

Qbo Robot as an Educational Assistant - Participants Feedback on the Engagement Level Achieved with a Robot in the Classroom

Raúl Madrigal Acuña[✉], Adrián Vega,
and Kryscia Ramírez-Benavides

Universidad de Costa Rica, San Pedro, Costa Rica
{Raul.Madrigal,Adrian.Vega,Kryscia.Ramirez}@ucr.ac.cr

Abstract. A non-humanoid robot is used to assist in an educational workshop of Quality Assurance and DevOps. The goal of this research was to determine the level of engagement shown by students of computer science in a presentation conducted by a University professor and assisted by a robot. The robot interaction was based on the Wizard of Oz technique. The order of actions between the professor and the robot was scripted and practiced before the workshop. After the workshop, a survey was conducted to assess the students' perception towards robot's shape, size, behavior and, performance. The survey also included the Godspeed Questionnaire Series to measure participant's perception of the robot and its effectiveness as an educational assistant. The results revealed the participants considered the robot featured personalized cognitive skills and exhibited an acceptable integration in the workshop.

Keywords: Non-humanoid social robot · Educational robot
Non-humanoid robot interaction · Wizard-of-Oz scenario

1 Introduction

We live in the technology era. Human labor has been gradually modified by machine operating forces that simplify and improve process efficiency. Robotics is one of the core concepts in technology. Robots are used as a problem solver in common daily situations. For this reason, different research has been done in many different fields such as health, education, and elderly care using robots [5, 9, 10].

Educational robotics has great potential as a learning tool at all levels, from kindergarten to the university. It provides rich opportunities for collaborative knowledge building and skills acquisition through the manipulation of and interaction with robots [9, 10]. However, we did not identify any case aiming the educational field using Qbo robots.

Many scientists in social robotics agree that the main requirements of a complex social interaction include communication, the recognition and expression of emotions, and some rudimentary form of personality. These features are widely thought to increase the believability of artificial agents and enhance the long-term engagement of

© Springer International Publishing AG, part of Springer Nature 2019
J. Chen (Ed.): AHFE 2018, AISC 784, pp. 163–171, 2019.
https://doi.org/10.1007/978-3-319-94346-6_16

people toward artificial companions. In addition, direct human-robot interactions analyzing humans' reactions also on the behavioral level are relatively rare [1, 6].

Similarly, a study used a non-humanoid robot as the interacting partner with infants of 2 to 3 years old. To test children's detection of contingency in a non-human agent in relatively natural settings. Further, by programmed recording, the actions performed by the robot for a participant in the contingent condition can be reproduced exactly for another participant in the non-contingent condition. Previous studies have suggested that infants attribute mental states to non-human objects including robots when the latter appear to interact with a person. Though there are some observational reports about the effect of behavioral contingency by a robot in group interaction, there have been few studies which systematically compare participants' reactions to contingent vs. non-contingent actions of a robot [2, 4].

In sum, to study if we can increment the level of engagement that a human shows to a particular activity using a robot. We needed a robot with the ability to communicate facial expressions while interacts using natural language with humans. Our focus on human-machine interaction allows us to use the Wizard-of-Oz (WOZ) technique to address the above-mentioned drawbacks. The underlying idea is that subjects are given the illusion that they are communicating with a computer system, while a human operator plays the role of the system [3, 7].

In the present experiment, we used a non-humanoid robot that interacts with people through facial gestures, natural language, and body movements. We predicted that the level of interest and engagement of the participants would increase if a robot is assisting an educational workshop. We designed the experiment pursuing to elapse the workshop in a natural way to the students. Therefore, the order of actions between the professor and the robot was scripted and practiced before the workshop.

2   Method

The main requirement to accomplish a successful Wizard-of-Oz scenario was to create a tool that allows us to control remotely the robot. To give the participants the illusion of the robot having the intelligence to control all its action by itself. This tool was developed as an Android Mobile application. Through the use of an interface in the mobile application we sent the actions to the robot such as head and body movements, facial gestures and text to speech. The robot received these actions using its built-in web API. The API is connected to all the robot sensors to enable a remote control of the robot through the web. The robot sent back to the application the streaming of the video cameras in its eyes.

2.1   Participants

Fifteen subjects took part in the experiment. All male with a Mean (M) of 24.4 years old and Standard Deviation (SD) of 3.66. All were computer science students and signed the informed consent to take part in the experiment that had been previously approved by the local ethical committee. The informed consent allows to analyze and publish results with a scientific purpose.[1]

2.2   Robot

The Qbo robot was developed by Thecorpora company. Thecorpora is a company that develops and promotes interactive robots at low costs to help people to learn more about robotics. The robot weights 10 kg approximately, 456 mm tall, has four infrared sensors, four ultrasonic sensors, two video cameras in the eyes, two microphones in the ears and one mouth. Qbo robot is equipped with Wi-Fi and Bluetooth, and it has administration panel software via web. It uses speech recognition and synthesis to interact with humans and can move its body using its two rear wheels and one front wheel [4]. Due to its shape is considered as a non-humanoid robot.

2.3   Environment

The experiment took place in a 30×30 m classroom on the last floor of a University building of three floors. The entire floor and classrooms were empty. The sits were aligned in a semicircle shape to distribute equally the vision and listening of the presenter and robot. Presenter and robot were placed in front and center of the semicircle.

2.4   Educational Workshop

It was a one hour workshop to present an Introduction to Jenkins for Continuous Integration and Continuous Delivery. A human speaker with the assistance of the Qbo robot led the presentation. The workshop took part in a university course of Quality Assurance, DevOps, and Agile Software Development.

2.5   Data Collection Artifact

A survey of five sections and forty-one questions in total was used to collect the participant's feedback. The survey includes the Godspeed Questionnaire Series (GQS). The GQS is one of the most frequently used questionnaires for assessing the success of robots in the field of Human-Robot Interaction. The GQS consists of five scales that are relevant to evaluate the perception of (social) Human-Robot Interaction. The scales are Anthropomorphism, Animacy, Likeability, Perceived Intelligence, and Perceived Safety [5]. From the five scales, we only focused on the analysis of the first four scales.

The first section of the survey asks for sit number, age, and gender. The second section was made based on the Godspeed Questionnaire to measure the robot perception of the participant. The questionnaire consists of five-point semantic differentials scale evaluating items such as "Fake—Natural". The third section asks about how the

---

[1] Mean (M) = Sum of values average.
  Standard Deviation (SD) = Estimated average of variation in a standardized set.
  Standard Error (SE) = Possible level of error in a measure.

participant felt during the workshop with the robot. The fourth section asks about participants' background regarding robotics. The last section asks about the opinion of the robot performance during the one hour workshop.

3 Procedure

The experiment was conducted in a wizard-of-oz scenario (the participants interact with the robot that they believe to be autonomous, but which is actually being operated by an unseen human being). The workshop used a presentation projected in a board to guide the participants in the content. In each of the slides, both the presenter and the robot followed a series of movements, facial gestures, and phrases that were previously scripted for the experiment. With the use of the mobile application, the unseeing human was controlling all the robot interactions during the workshop.

Before the experiment, the participants were informed that they were going to interact with a robot. No information regarding how the robot was controlled was informed to the participants. The experiment lasted approximately 45 min. After the experiment, the participants filled out the survey what took about 10–15 min to complete.

4 Results

We analyzed the data from the participant's responses to the survey. The results were separated into three aspects. For the first aspect, we used the section number four of the survey to analyze the background of the participants regarding robotics. The second aspect, covers the results from section number five of the survey, to analyze the robot's performance during the workshop. And for the last aspect, we used the first three sections of the survey to analyze the results of the Godspeed Questionnaire Series.

4.1   Participants' Background Regarding Robotics

The next three questions were used to retrieve the information and the results are shown in Fig. 1.

1. How would you describe your relationship with robots?
2. How would you describe your contact on a daily basis with robots?
3. How would you describe your previous experience with non-humanoid robots?

From Fig. 1 we can determine that the participants are showing almost no contact with robots on a daily basis. Having a Mean of 2.67, a Standard Deviation of 1.29 and Standard Error of .33. Although, most of the participants showed interest in robotics. Having a Mean of 3.73, a Standard Deviation of .70 and Standard Error of .18. And even a less number of the participants have had previous contact with non-humanoid robots. Having a Mean of 2.47, a Standard Deviation of 1.25 and Standard Error of .32.

Additionally, the survey asks for the means in which the participants have been getting information about robotics. From the results, we identified the most common concepts. An 80% of the participants voted for internet as the main channel to get

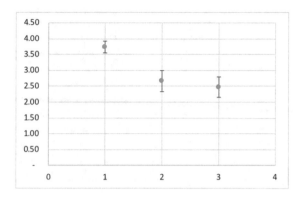

Fig. 1. Shows the mean, standard deviation and standard error of the answers for the three questions related to participant's background and previous experience with robots.

information. A 60% voted for television, a 40% voted for books and 13% voted for none mean. To categorize the type of information they were getting, they had to select between popular as movies, news or tv shows and academic as papers, professional talks or university courses to relate the content with a specific category. An 80% of the content was categorized as popular and a 73% of the content as academic.

4.2   Qbo Robot Performance

The participants had to select a value between five options for each question. Having the most left option as the minimum, worst or less value, the middle one as the normal or regular option and the most right option as the maximum, best or highest value to describe what they think about the question. Next are the rating questions we used to retrieve the information and the results are shown in Fig. 2.

1. Rate the voice volume.
2. Rate the voice quality.
3. Rate the speech speed.
4. Rate the physical distance between your sit and the robot position.
5. Rate the angle of vision to the robot.
6. Rate the robot's size.

From Fig. 2, we can determine the robot's performance during the workshop. Starting with the Voice Volume Mean (M = 2.73, SD = 0.7, SE = .18) under the "normal" value, because 40% of the participants qualified this feature as "low". The voice quality Mean is normal (M = 3.0, SD = .65, SE = .17). The speech speed Mean (M = 3.73, SD = .8, SE = .21) is above the normal, because 55% of the participants qualified as "fast" or "too fast" while is speaking. The physical distance between the robot and each sit of the participants, shows a normal Mean (M = 2.93, SD = .59, SE = .15). The angle of vision for the participant to the robot shows a normal mean (M = 3, SD = 1, SE = .26). Lastly, the robot size Mean (M = 2.47, SD = .52, SE = .13) is under the normal, because 53% qualified the robot as small.

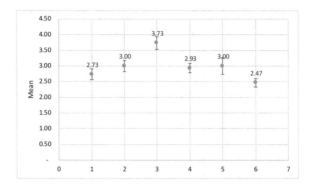

Fig. 2. Shows the mean, standard deviation and standard error of the answers for the six questions related the qualification to robot's performance during the workshop.

4.3   Godspeed Questionnaire Series

Anthropomorphism, Animacy, Likeability and Perceived Intelligence were the scales we analyzed from the questionnaire. For each of the scales, there is a five-point semantic differential. The internal consistency of the scales was evaluated based on the Cronbach's Coefficient Alpha [8].

The results indicate a good internal consistency with values for animacy with 0.84, anthropomorphism with 0.77, likeability with 0.91 and perceived intelligence with

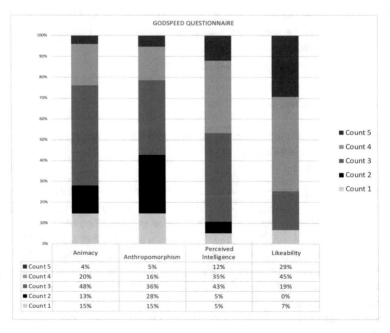

Fig. 3. Frequencies analysis of the data collected from the participant's responses to the survey.

0.69. The last can be considered an acceptable value but it is worthy to reevaluate it in a future research. These values can be seen in the frequencies analysis of Fig. 3. The highest values are for likeability and perceived intelligence with a 74% and 47% of the data above the medium value. Different from, animacy and anthropomorphism with values of 24% and 21% above the median value.

To demonstrate the results obtained in Fig. 3. we now present in more detail the values for each of the semantic differentials. The values are in descendent order and show how the Qbo robot was perceived and described for the participants. With attention to Fig. 4, the robot was perceived as 80% polite, 80% cute, 79% like, 77% nice and 75% friendly.

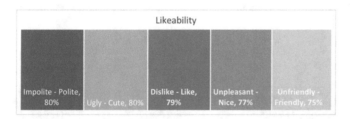

Fig. 4. Semantic differentials result for likeability scale of the Godspeed Questionnaire Series.

The results for the perceived intelligence can be seen in Fig. 5. Where the robot was described as 79% cult, 71% responsible, 69% competent, 63% intelligent and 61% reasonable.

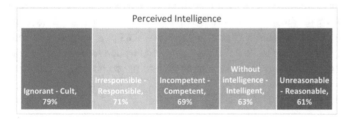

Fig. 5. Semantic differentials result for the perceived intelligence scale of the Godspeed Questionnaire Series.

Without surprises, the results for the animacy scale where close the medium value, bearing in mind the robot has a non-humanoid aspect. As it can be seen in Fig. 6 the robot was described as 67% active, 65% interactive, 55% alive, 51% organic and 48% human.

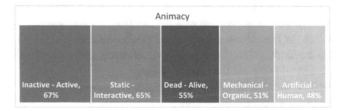

Fig. 6. Semantic differentials result for the animacy scale of the Godspeed Questionnaire Series.

Ultimately, the results for the anthropomorphism indicates the robot as 64% conscious, 56% of fluid movements, 55% natural, 48% alive and 47% of a human aspect (Fig. 7).

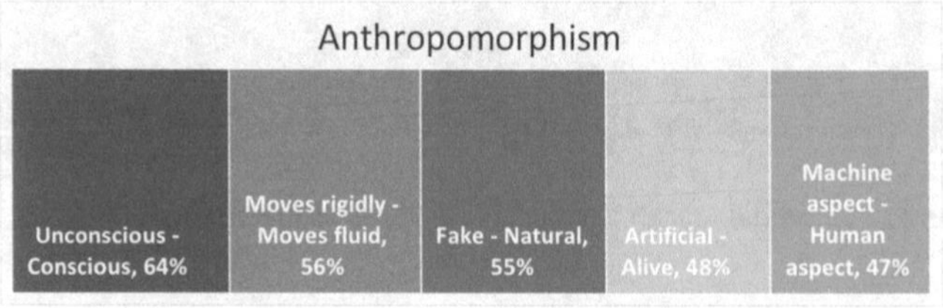

Fig. 7. Semantic differentials result for the anthropomorphism scale of the Godspeed Questionnaire Series.

In a general way, it can be said that likeability was the most remarkable feature of the interaction with the robot. Where it is described especially as polite, cute and likely. Followed by perceived intelligence being perceived as cult, responsible and competent. Subsequently, according to its animation is valued as active, interactive and alive. And finally, the anthropomorphism is characterized as conscious, of fluid and natural movements.

5   Discussion

In general, the Wizard-of-Oz technique was accomplished and gave us a way to addressed the experiment. The participants did not notice that the robot was controlled by a human. In fact, they thought the robot had cognitive skills during the interaction over the workshop presentation. This effectiveness is presented in the results of the Godspeed Questionnaire, where the participant described the robot using high values of sympathy, intelligence, animosity, and anthropomorphic, in that order.

The results also revealed that the participants considered the integration between the robot and activity as acceptable. Achieving positive scores regarding the information presented, group size, setup distance between robot and group, the visual sight to the robot, and robot voice tone. The participants also pointed out the utility of using robots in other activities such as: as a personal assistant, an educator or as a supermarket helper. The successful interaction between Qbo and the participants suggest the usability of these robots in different applications and environments. For future evaluations, it is required to evaluate why the participants did not notice a difference in the level of engagement by including the Qbo robot in the workshop. We must also consider other limitations evaluated by the participants such as the robot's size, voice volume and speech speed.

Finally, this research is a pioneer in the evaluation of the Qbo Robot under these circumstances, so it is required continue with the evaluation of this kind of robot in different environments using similar tests to discriminate more accurately the effect of the robot physical specs, personalization, and role activities.

Acknowledgments. This work was partially supported by Centro de Investigaciones en Tecnologías de la Información y Comunicación (CITIC), Escuela de Ciencias de la Computación e Informática (ECCI) both at Universidad de Costa Rica (UCR). Grand No. 834-B7-267. We would like to thank Programa de Posgrado en Computación e Informática and Sistema de Estudios de Posgrado at UCR for their support. Additionally, thanks to the User Interaction Group (USING) for providing ideas to refine and complete the research.

References

1. Lakatos, G., Gacsi, M., Bereczky, B.: Emotion attribution to a non-humanoid robot in different social situations. Plos One, **9**(12), e114207 (2015)
2. Yamamoto, K., Tanaka, S., Kobayashi, H., Kozima, H., Hashiya, K.: A non-humanoid robot in the "Uncanny Valley": experimental analysis of the reaction to behavioral contingency in 2–3 year old children. Plos One **4**(9), e6974 (2009)
3. Gnjatović, M., Dietmar, R.: Gathering corpora of affected speech in human machine interaction: refinement of the Wizard-Of-Oz technique. In: Proceedings of the International Symposium on Linguistic Patterns in Spontaneous Speech (2006)
4. Artificial Intelligence at Home - Qbo Robot. http://thecorpora.com/qbo-robot/
5. Weiss, A., Bartneck, C.: Meta analysis of the usage of the godspeed questionnaire series. In: Proceedings of the IEEE International Symposium on Robot and Human Interactive Communication (RO-MAN2015), Kobe, pp. 381–388 (2015)
6. Dubal, S., Foucher, A., Jouvent, R., Nadel, J.: Human brain spots emotion in non humanoid robots. Social Cognitive and Affective Neuroscience, vol. 6 (2011)
7. Riek, L.D.: Wizard of Oz Studies in HRI: A Systematic Review and New Reporting Guidelines, vol. 1. University of Notre Dame (2012)
8. Cronbach, L.J.: Coefficient alpha and the internal structure of tests. University of Illinois, Springer (1951)
9. Baxter, P., Ashurst, E., Read, R., Kennedy, J., Belpaeme, T.: Robot education peers in a situated primary school study: personalisation promotes child learning. Plos One **12**(5), e0178126 (2017)
10. Bravo, F.A., González, A., González E.: Interactive Drama With Robots for Teaching Non-Technical Subjects, Pontificia Universidad Javeriana (2017)

Author Index

© Springer International Publishing AG, part of Springer Nature 2019
J. Chen (Ed.): AHFE 2018, AISC 784, pp. 173–174, 2019.
https://doi.org/10.1007/978-3-319-94346-6

Printed in the United States
By Bookmasters